CALCULUS
in a new key

D.L. ORTH

APL PRESS
Swarthmore, PA

ISBN 0-917326-05-9

Preface

My purpose in writing this text was to present a substantial mathematical treatment of derivatives and integrals as tools for analyzing and constructing functions, as opposed to a formal treatment of limits. In developing this treatment I have stressed that many analytical tools can themselves be described as explicit functions. Consequently, once derived, these tools remain readily available for use, they can be combined in the usual ways for functions to produce new tools, and they too can be analyzed in terms of other functions.

The notation used in this text is APL. The precision of APL permits each expression to be analyzed with confidence, and the analysis can be carried out with the aid of an APL computer if available. Many functions have particularly simple descriptions in APL, as compared to their formulations in conventional notation. Thus the use of APL has permitted selection from a wider range of topics than one finds available with conventional notation, and has allowed greater reliance on exercises.

Polynomials are emphasized early and used in the construction of other functions. Every function is introduced by way of at least one explicit construction. For example, the exponential function is defined as the limit of a family of polynomials, and inverse functions are constructed in terms of a rootfinder. Taylor polynomials and the construction of solutions of initial value problems are emphasized.

The limit concept is used in an informal way, although the precise definition underlies the estimate of tails of families of polynomials and of the proof of the existence of the definite integral of a monotonic function. Continuous functions are not emphasized, and the definite integral is developed for functions of bounded variation. No formal notation for the derivative and integral operators is used. Instead, an informal naming convention has been adopted, where the derivative of a monadic function *F* is the monadic function named *D̲F* and the definite integral of *F* is the dyadic function named *S̲F*. Thus the question of the existence of the derivative and definite integral of *F* becomes one of defining the domains of *D̲F* and *S̲F*. Techniques of integration are treated but not emphasized.

The rudiments of matrix algebra are introduced early, with emphasis on solutions of linear equations and the representation of linear functions in terms of the matrix product. The results are applied to fitting polynomials, quotients of polynomials, numerical integration, and integration of rational functions.

The text material can be rearranged in several ways; for example, with the exception of the algebra of Taylor polynomials in Section 4.5, Chapter 1 can be followed immediately by Section 3.1 and Chapter 4. However, some care must be taken with those exercises which depend on preceding chapters.

It is also possible to vary the depth to which material is initially developed. For example, it is easy for students to understand the effects of the functions on polynomials defined in Chapter 3, so that it is not necessary to include detailed discussions of their definitions (Sections 3.2-3.5) in a first pass through the chapter.

I am pleased to acknowledge the assistance of Professor James England of Swarthmore College, who has used the text in his freshman calculus class, and Mr. E.E. McDonnell of IBM, and Professor Charles Sims of Rutgers University, who carefully read preliminary drafts. Their corrections and comments have been most valuable.

I would also thank Ms. Agnes Carlin and Ms. Mary Flowers for their excellent work in typing the manuscript, and the IBM Corporation for its support of this project. I am indebted to Ms. Elizabeth Llanso for the development of the APL Text Editor, on which this manuscript was prepared.

A special debt is owed Dr. K. E. Iverson of IBM, who has been most generous in providing me with his mathematical expertise and insights into the problems of teaching mathematics.

D. Orth

Trenton, New Jersey
June, 1976

Contents

0
Introduction

Generally speaking, the calculus is concerned with two fundamental techniques for analyzing and constructing functions, one based on differences of function values of neighboring arguments and the other on sums of function values. For each function of one argument there are two associated functions, called the <u>derivative</u> and the <u>definite integral</u> of the function. Rates at which values of a function change are related to values of its derivative. Averages of values of a function are related to values of its definite integral.

Fundamental concepts in many areas of applied mathematics are represented by derivatives and definite integrals; this accounts for the wide range of applications of the calculus. For example, if a function represents the position of a moving object, then its derivative represents the velocity of the object. Also, if a function represents the pressure at points on a side of a tank filled with water, then its definite integral is related to the force exerted by the water on the side of the tank.

The present text treats the analytical techniques of the calculus and several representative applications. It differs from most calculus texts in two major respects. Firstly, the text is so organized that whenever a function is introduced it can be defined explicitly in terms of previously defined functions. Thus at the beginning we concentrate mainly on polynomials, which are defined in terms of addition, multiplication, and powers. Polynomials are then used to define other functions such as growth functions and circular functions.

Secondly, the notation used in this text is that used in Iverson [1,2] and commonly known as APL. The simplicity and precision of this notation permits an algorithmic treatment of the calculus. In particular, every expression in the text can be executed by simply typing it on an appropriate computer keyboard. While the use of the computer is not absolutely essential to this course, it can be an effective tool for experimentation. Students should repeat the examples in the text, create new examples, and experiment with the exercises, and these things can be done quickly and accurately on a computer.

The simplicity and precision of the notation also permits a greater burden than is usual to be placed on the exercises. Thus the text often presents only a concise overall description and leaves it to the exercises to bring out the details or to develop related ideas. The serious student may therefore participate in the development of the calculus, but the casual reader may encounter difficulties. The point at which a set of exercises can be attempted is indicated by the marginal symbol ⌹ followed by the number of the first exercise in the group. The groups of exercises are separated by lines.

A summary of the notation used in this text is presented below. A small amount of notation is introduced later in the text as needed. Any student with the proper mathematical background for the study of the calculus who is not familiar with the notation should find this summary an adequate preparation. Students with access to an appropriate computer will find it a convenient tool for experimenting with the notation. The relation between this notation and conventional notation and the consequences of the differences are discussed in Appendix A of Iverson [1].

0.1 NOTATION

Each function has a symbol (unlike the power function in conventional notation), the symbol is unique (unlike the use of ÷ and / for division) and the symbol must be used (unlike the elision of the multiplication sign in expressions like (*X*+1)(*X*+2)). For example:

```
      3+2
5
      3×2
6
      3⋆2                              (the power function)
9
      2⋆3
8
```

There is one form for <u>dyadic functions</u>, that is, functions of two arguments: the symbol occurs between the arguments, as in *X*+*Y* and *X*⋆*Y*.

There is one form for all <u>monadic functions</u>, that is, functions of one argument: the symbol occurs before the argument, as in -*X* and *SIN T*.

Each symbol is commonly used in two distinct ways, to denote both a dyadic function and a (usually related) monadic function. This practice is familiar from the use of the minus sign to denote the dyadic function of subtraction (as in *A*-*B*) and the monadic function of negation (as in -*B*). For example, the dyadic function denoted by × is multiplication, while the monadic function denoted by × is called <u>signum</u>. The value ×*X* is 1 if *X* is positive, negative 1 if *X* is negative, and 0 if *X* is 0.

Constants are written in the familiar decimal form except that negative numbers are denoted by a raised minus sign as in ¯3.14. This symbol does not denote a function but, like the decimal point, is part of the representation of the number. Exponential notation is also used, as in 314*E*¯2 and .314*E*1 for 3.14 and ¯314*E*¯2 for ¯3.14. For example:

```
      6÷2
3
      ÷2
0.5
      -4
¯4
      |¯3.1                           (the magnitude function)
3.1
      |7
7
      4⌈8                             (the maximum function)
8
      4⌊8                             (the minimum function)
4
```

The symbols < ≤ = ≠ ≥ > are used for relations and denote functions rather than assertions. For example, the value of 3<5 is 1 because the relation holds and the value of 5<3 is 0 because the relation does not hold. For example:

```
      3≥4
0
      3≥3
1
      3≠3
0                                                              ⍠1
```

The symbols ∧ (for and) and ∨ (for or) denote logical functions; the arguments of these functions must have values 0 or 1. For example:

```
      1∧1                             1∨1
1                               1
      1∧0                             1∨0
0                               1
      0∧1                             0∨1
0                               1
      0∧0                             0∨0
0                               0
```

The symbol ~ (for not) denotes the monadic function whose arguments are 0 and 1 and whose values ~0 and ~1 are 1 and 0 respectively.

The symbol ← denotes assignment of a value to a variable name. For example, *A*←1 may be read as "let *A* be

1" or as "*A* is 1". The use of the equal sign for this purpose, as in "let *A*=1", is avoided to prevent conflict with its use as a relation. For example:

```
      A←1                               B←3+7
      A                                 B
1                              10
```

The order in which the functions in an expression are executed is determined by parentheses in the usual way, but in the absence of parentheses the order is governed by a simple rule, which can be expressed in two equivalent ways:

> Every function takes as its right argument the value of the entire expression to the right of it;
>
> The functions are executed in order from right to left.

For example, the expression

```
      8+4*P≤|Q
```

is equivalent to the fully parenthesized expression

```
      8+(4*(P≤(|Q)))
```
⌹2

Functions are extended to arrays in a systematic manner. Arrays are lists (also called vectors), tables (also called matrices) and higher dimensional arrays such as lists of tables, tables of tables, etc. For example:

```
      1 4 ¯3 + 0 6 3
1 10 0

      X←1 4 ¯3
      Y←0 6 3
      X×Y
0 24 ¯9
```
⌹3

```
      0 1 2 ∘.+ 5 3 ¯2
5 3 ¯2
6 4 ¯1
7 5  0
```

The symbols ∘. in the last example indicate that the result is to be a table obtained by applying the indicated function (in this case, +) to each pair of elements (the numbers in an array are called elements) taken one from the first and one from the second. The symbols ∘. can be used in conjunction with any dyadic function such as × or ⌈. ⌹4

Expressions such as *A*+*B* and *A*×*B* normally apply only if *A* and *B* have the same shape. That is, if *A* is a vector then *B* is also a vector with the same number of elements, or if *A* is a matrix then *B* is also a matrix with the same number of rows and columns. An exception is made if one of the arguments consists of a single element, in which case it is applied to each element of the other argument. For example:

```
      3 + 4 7 ¯3
7 10 0
      A←2
      B←1 2 3
      A*B                            B*A
2 4 8                          1 4 9                     ⌹6
```

Numbers, which are also arrays, will be called *scalars*, and functions whose arguments are scalars will be called *scalar functions*. In reference to the extensions of functions discussed in the preceding paragraphs, we say that *scalar functions extend element-by-element to arrays*.

Elements can be selected from arrays by various functions. The simplest are the *take* function and the *drop* function which are denoted by ↑ and ↓ respectively. Their definitions should be clear from the following examples:

```
      3↑1 2 3 4 5                    3↓1 2 3 4 5
1 2 3                          4 5
      ¯3↑1 2 3 4 5                   ¯3↓1 2 3 4 5
3 4 5                          1 2
      7↑1 2 3 4 5                    ¯7↑1 2 3 4 5
1 2 3 4 5 0 0                  0 0 1 2 3 4 5

      M←1 2 3∘.+1 2 3
      M
2 3 4
3 4 5
4 5 6
      2 3↑M                          3 2↑M
2 3 4                          2 3
3 4 5                          3 4
                               4 5

      ¯1 3↑M                         ¯1 ¯1↓M
4 5 6                          2 3
                               3 4

      2 0↓M                          0 ¯1↓M
4 5 6                          2 3
                               3 4
                               5 6
```

The selection function denoted by the slash (/) is called *compression*. For example:

```
      1 0 1 1 0 1/2 3 4 5 6 7
2 4 5 7
      1 1 0 0 1/8 7 6 5 4
8 7 4
```

The dyadic function denoted by reverse slash (\) is called expansion. For example:

```
      1 0 1 1 0 1\2 4 5 7
2 0 4 5 0 7
      1 1 0 0 1\8 7 4
8 7 0 0 4
```

More general selection is provided by indexing. For example, if *Y*←5 6 7 8 then *Y*[1] is 5 and *Y*[4 2] is the vector 8 6. In this example the first index is 1 and it is therefore referred to as 1-origin indexing. In 0-origin indexing the first index is 0. In 0-origin indexing *Y*[0] is 5 and *Y*[3 1] is the vector 8 6.

The index origin in use is indicated formally by specifying the value of the index origin denoted by ⎕*IO*. Thus ⎕*IO*←1 indicates that 1-origin indexing will be in effect until some later specification of ⎕*IO*, and ⎕*IO*←0 indicates that 0-origin indexing is in use. In this text, 0-origin indexing will be used throughout.

Indexing also applies to tables. For example, if

```
      T←1 5∘.+1 2 3
      T
2 3 4
6 7 8
```

then *T*[1;2] is the element at the intersection of row 1 and column 2 of *T*. That is:

```
      T[1;2]
8
```

Also, *T*[0;] is row 0 of *T* and *T*[;2] is column 2 of *T*. Thus:

```
      T[0;]                          T[;2]
2 3 4                          4 8
```

⌹7

Catenation is a dyadic function denoted by the comma; it serves to chain two arrays together. For example:

```
      X←1 2
      Y←5 6 7
      X,Y                       X,0
1 2 5 6 7             1 2 0
      Y,X                       0,X
5 6 7 1 2             0 1 2

      A←1 3∘.×2 0 1
      A
2 0 1
6 0 3
```

```
      A,X                  X,A
2 0 1 1               1 2 0 1
6 0 3 2               2 6 0 3
```
⌹9

Summation of the elements of a vector *X* is denoted by +/*X*. For example:

```
      +/1 2 3 4
10
```

The result is equivalent to that obtained by writing the function symbol between each pair of successive elements in the vector, as in 1+2+3+4. This applies to any scalar dyadic function and is called reduction by that function. For example:

```
      ×/1 2 3 4
24
      ⌈/1 2 3 4
4
```

If *V* is a vector, then the vectors 1↑*V* and 2↑*V* and 3↑*V*, etc., are called prefixes of *V*; scan is denoted by a reverse slash (\) and produces the reduction of all prefixes of a vector argument. For example:

```
      +\1 5 2 4
1 6 8 12
      ×\1 5 2 4
1 5 10 40
      ⌈\1 5 2 4
1 5 5 5
```
⌹10

If *M* is a matrix then +/*M* produces the sum along the rows and +⌿*M* produces the sum along the columns. For example:

```
      M← 0 3∘.+1 2 3
      M
1 2 3
4 5 6
      +/M                  +⌿M
6 15                  5 7 9
```
⌹11

The monadic function denoted by ⍳ (the Greek letter iota) and called the index generator yields, when applied to a non-negative integer *N*, the vector of indices of a vector of *N* elements. For example:

```
      ⍳3                   ⍳5
0 1 2                 0 1 2 3 4
```

The monadic function denoted by ⍴ (the Greek letter rho) and called the size function produces the size of the argument array. The size of a vector is the number of elements of the vector. For example:

```
      ρ1 3 7 9
4
      ρι5
5
      ιρ 1 2 7 9
0 1 2 3
```

The size of a matrix is a two element vector consisting of the number of rows and the number of columns of the matrix. For example:

```
      M←1 2∘.× 2 3 4
      M
2 3 4
4 6 8
      ρM
2 3                                                          ⌹13
```

The dyadic function denoted by ρ is called reshape and reshapes the right argument to the size specified by the left argument. For example:

```
      ι9
0 1 2 3 4 5 6 7 8
      3 3ρι9
0 1 2
3 4 5
6 7 8
      5ρι9
0 1 2 3 4
      7ρ1 2 3
1 2 3 1 2 3 1
      4ρ2
2 2 2 2                                                      ⌹14
```

There are several functions which rearrange the elements in arrays, and one is denoted by ⌽. The monadic function ⌽ is called reverse and the dyadic function ⌽ is called rotate; the definitions, as they apply to vectors, should be clear from the following examples:

```
      ⌽ 1 2 3 4 5
5 4 3 2 1
      1 ⌽ 1 2 3 4 5                    ¯1⌽ 1 2 3 4 5
2 3 4 5 1                         5 1 2 3 4
      2 ⌽ 1 2 3 4 5                    ¯2 ⌽ 1 2 3 4 5
3 4 5 1 2                         4 5 1 2 3                  ⌹16
```

0.2 EXPRESSIONS, IDENTITIES, and PROOFS

An algebraic <u>expression</u> is a well-formed sentence, that is, a sentence which can be executed if suitable values are assigned to each of its variables. For example, $3+X$ is an expression but 3+ is not.

Two expressions are said to be <u>equivalent</u> if they produce the same results upon any substitution of the same values for variables. Equivalence of expressions will be indicated by writing one of the expressions directly below the other. Two equivalent expressions so written will be called an <u>identity</u>. For example, the identity

$X+Y$
$Y+X$

expresses the commutativity of +.

A sequence of equivalent expressions will be called a <u>sequence of identities</u>. For example:

$X+(Y+Z)$
$(X+Y)+Z$
$(Y+X)+Z$

Since any pair of consecutive expressions in a sequence of identities are equivalent, all expressions in the sequence are equivalent. In particular, the first and last expressions in the sequence are equivalent. If one indicates why each expression in a sequence is equivalent to its predecessor, the result is a proof of the equivalence of the first and last expressions. For example:

$X+(Y+Z)$	
$(X+Y)+Z$	associativity of +
$(Y+X)+Z$	commutativity of +

It is often useful to label expressions and identities for future reference. Labels will be applied as follows:

$X+(Y+Z)$ [10.4.2]
$(X+Y)+Z$

The 10 and 4 in 10.4.2 refer to the chapter and section respectively. Within each section figures, tables, expressions and identities are all labelled with one sequence.

Many exercises are concerned with establishing the equivalence of sequences of expressions. It is good practice to <u>illuminate</u> each by working through examples with particular values assigned to the variables. This is also true of the proofs presented in the text. ⌹17

0.3 FUNCTION DEFINITION

The functions for which the notation provides symbols, such as +, ×, ⌽, etc., are called primary functions. It is possible to define and name other functions, called secondary functions, that can then be used as new primary functions.

We will now present a scheme for defining functions which was first used in Iverson [2] and is called the αω form of function definition. It is not the canonical form of APL function definition commonly used on a computer, but can be used on a computer if the appropriate defined functions are available.

One set of such functions is listed in Appendix A. If they are entered on an APL computer, then one may define a function as follows. First enter *DEF* (followed by a carriage return) and then enter the function definition in the αω form. The result will be to fix the definition of the function. For example:

```
      DEF
POW:α*ω
      2 POW 3
8
```

A function definition can be displayed by entering its name within the function *DEF*. For example:

```
      DEF
POW
POW:α*ω
```

A monadic function is defined by an expression in which the argument is represented by either α (the Greek letter alpha) or ω (the Greek letter omega). For example, the expressions ω*2, α×2, A×ω, ÷/α*2 define monadic functions.

A defined function is given a name when its defining expression is preceded by the name and a colon. For example, *SQ*:ω*2 is read "the function named *SQ* is defined by the expression ω*2". The function *SQ* can be used as a new primary function. For example:

```
      SQ 1 2 3
1 4 9
      3×SQ 1 2 3
3 12 27
```

Also, if *F*:ω+3×*SQ* ω then

```
      F 5                    +/F 3 5
80                     110
```

A dyadic function is defined by an expression in which the left argument is represented by α and the right argument by ω. For example, if *POW*:α*ω then

```
      2 POW 3
8
```

```
      3 POW 2
9
```

The most general αω form of function definition is represented schematically as follows:

name:first expression:proposition:second expression

A <u>proposition</u> is an expression whose values are either 0 or 1. For example, the expression ∧/ω>0 represents a proposition whose value is 1 if all elements of an argument vector are positive and 0 otherwise. The value of a function whose definiton can be represented schematically as above is the value of the first expression if the value of the proposition is 0 and the value of the second expression if the value of the proposition is 1. The proposition is evaluated before either of the expressions. For example, if

```
      F:1↓ω : 3<⍴ω : ¯1↓ω
```

then

```
      F 2 5
5
      F 2 5 7
5 7
      F 2 5 7 9
2 5 7
```

```
      1↓2 5
5
      1↓2 5 7
5 7
      ¯1↓2 5 7 9
2 5 7
```

As another example, consider the <u>recursively defined</u> factorial function

```
      FAC:ω×FAC ω-1 : 0=ω : 1
```

The function *FAC* is said to be recursively defined because it occurs in its own definition. For example, to produce the value *FAC* 3, we see that the value of the proposition 0=ω is 0 for the argument 3, so that *FAC* 3 is 3×*FAC* 2 (that is, the value of the first expression in the definition of *FAC*). Similarly, *FAC* 2 is 2×*FAC* 1 and *FAC* 1 is 1×*FAC* 0. The value of the proposition 0=ω is 1 for the argument 0, and so *FAC* 1 is 1 (that is, the value of the second expression in the definition of *FAC*). Thus *FAC* 1 is 1×1, or 1, *FAC* 2 is 2×1, or 2, and *FAC* 3 is 3×2, or 6. ⌹21

The names that can be given to functions and variables consist of any number of alphabetic characters (including underscored letters such as <u>*A*</u>) and numeric characters (digits 0 through 9); the first character must be alphabetic.

In this text we will adopt the following convention: all names appearing in Appendix A have the suffix 9; consequently, to avoid conflicts, we will not use the suffix 9 in function and variable names.

0.4 EXAMPLES

The examples in the text appear as they would when executed at an appropriate computer keyboard. When an expression is entered at a keyboard, followed by a carriage return, its value is displayed unless the function last executed is assignment. For example:

```
      X←1 0 1/3 4 5
      1 0 1/3 4 5
3 5
```

The value of any expression is displayed if, when entered, it is preceded by the symbols ⎕←. For example:

```
      ⎕←X←1 0 1/3 4 5
3 5
      ⎕←1 0 1/3 4 5
3 5
```

The dyadic format function is used to display a result compactly so as to include a specified number of decimal digits. For example:

```
      A←3 4.1 5.23
      0⍕A
3 4 5
      1⍕A
3.0 4.1 5.2
      2⍕A
3.00 4.10 5.23
```

1 The Derivative

1.1 DIFFERENCES AND SUMS

The result of the function *DIFF*:(1↓α)-¯1↓α is called the vector of <u>pairwise differences</u> of successive elements of its argument. For example:

```
      F←8 ¯2 1 6 ¯6 ¯6

      DIFF F
¯10 3 5 ¯12 0
      F[3]-F[2]
5
      (DIFF F)[2]
5
```
⌹1

The function *SUM*:0,+\ω is called the <u>augmented plus scan</u> because it produces the identity element of + appended to the result of the function +\ω. For example:

```
      F
8 ¯2 1 6 ¯6 ¯6

      SUM F                      +\F
0 8 6 7 13 7 1             8 6 7 13 7 1
```

Zero is called the identity element of + because *X*+0 and 0+*X* equal *X*. ⌹7

SUM is, roughly speaking, the inverse function of *DIFF*. Precisely, for vectors *F* and scalars *C* we have the identities

```
      F                    F                          [1.1.1]
      DIFF C+SUM F         F[0]+SUM DIFF F
```

For example:

```
      F
8 ¯2 1 6 ¯6 ¯6

      DIFF 5+SUM F              F[0]+SUM DIFF F
8 ¯2 1 6 ¯6 ¯6           8 ¯2 1 6 ¯6 ¯6
```

The proofs of Identities 1.1.1 will be discussed in the exercises, as will elementary applications of *SUM* and *DIFF*. These functions are of fundamental importance to the study of the calculus and will be used throughout the text. In particular, the function *DIFF F ω* is closely related to the derivative of *F*, while *SUM F ω* is closely related to its definite integral. Identities 1.1.1 reflect the relationships between derivatives and definite integrals, which are precisely described by the Fundamental Theorems of the Calculus (Chapter 5). ⍠10

1.2 GRAPH SKETCHING

An important aspect of analyzing a function is sketching its graph, and an obvious strategy for producing a graph is to plot many points on the graph. A more interesting and efficient strategy is to produce the graph from a few points. This strategy is important for two reasons other than efficiency. Firstly, it forces us to look closely at the process of graph sketching in order to isolate those characteristics of functions that are relevant to their graphs. Secondly, it is not always possible to determine a large number of values of a function and consequently its graph must be produced from a few values.

As an example of a strategy for producing a graph from a few function values consider the function *F*:¯2+(¯1×*ω*)+*ω*⋆2. Let *X*←¯2 1 3.5 and choose a vector *S* whose elements have values close to zero, say *S*←0.2 0.1 ¯0.2. Now plot the points with coordinates *X*[*I*],*F X*[*I*] and (*X*+*S*)[*I*],*F* (*X*+*S*)[*I*] when *I* is 0, 1 and 2. Next, sketch segments of the graph connecting each pair of neighboring points as in Figure 1.2.1. These segments approximate the local behavior of the graph.

A sketch of the entire graph is then produced by connecting the segments with a smooth curve, as in Figure 1.2.2.

The segments in Figure 1.2.1 are approximately straight line segments, and these line segments are segments of the secant lines passing through the coordinate points *X*[*I*],*F X*[*I*] and (*X*+*S*)[*I*],*F* (*X*+*S*)[*I*]. The slopes of these secant lines can be produced by the difference-quotient function *QF*:((*F ω*+*α*)-*F ω*)÷*α*.

```
      S QF X
¯4.8 1.1 5.8
```
⍠35

1.3 DIFFERENCE-QUOTIENTS

With each monadic scalar function *F* there is associated a dyadic function

QF:((*F ω*+*α*)-*F ω*)÷*α*

called the difference-quotient of *F*. As we saw in the preceding section, values *S QF X* for non-zero scalars *S* are slopes of secant lines joining the points *X*,*F X* and (*X*+*S*),*F X*+*S*. In this section we will concentrate on formal

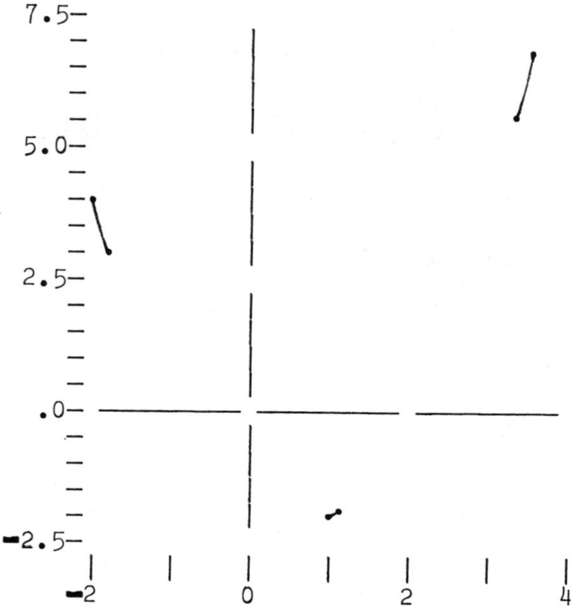

SEGMENTS OF A GRAPH
FIGURE 1.2.1

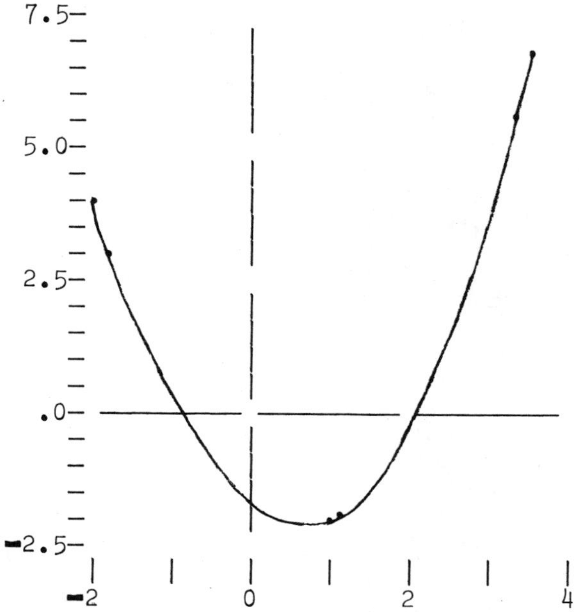

GRAPH PRODUCED BY CONNECTING SEGMENTS
FIGURE 1.2.2

identities for difference-quotients, returning to the geometric interpretation in Section 1.4.

If $\underline{Q}F$ and $\underline{Q}G$ are the difference-quotients of F and G respectively, then the difference-quotient of the sum function $(F\ \omega)+G\ \omega$ can be defined directly in terms of $\underline{Q}F$ and $\underline{Q}G$. This is also true of the product function $(F\ \omega)\times G\ \omega$, the reciprocal function $\div F\ \omega$, and the composite function $F\ G\ \omega$. Thus the difference-quotient of a function defined by a complicated expression can be analyzed in terms of the difference-quotients of its constituents. That is, we have the following rules for difference-quotients:

Plus Rule for Difference-Quotients. If $P\ \omega$ is equivalent to $(F\ \omega)+G\ \omega$, then

$$\alpha\ \underline{Q}P\ \omega$$
$$(\alpha\ \underline{Q}F\ \omega)+\alpha\ \underline{Q}G\ \omega$$

Times Rule for Difference-Quotients. If $T\ \omega$ is equivalent to $(F\ \omega)\times G\ \omega$, then

$$\alpha\ \underline{Q}T\ \omega$$
$$((\alpha\ \underline{Q}F\ \omega)\times G\ \omega)+((F\ \omega)\times\alpha\ \underline{Q}G\ \omega)+\alpha\times(\alpha\ \underline{Q}F\ \omega)\times\alpha\ \underline{Q}G\ \omega$$

Reciprocal Rule for Difference-Quotients. If $R\ \omega$ is equivalent to $\div F\ \omega$, then

$$\alpha\ \underline{Q}R\ \omega$$
$$-(\alpha\ \underline{Q}F\ \omega)\div((F\ \omega)\star 2)\times 1+\alpha\times(\alpha\ \underline{Q}F\ \omega)\div F\ \omega$$

Composite Rule for Difference-Quotients. If $C\ \omega$ is equivalent to $F\ G\ \omega$, then $S\ \underline{Q}C\ X$ is 0 if $S\ \underline{Q}G\ X$ is 0, and otherwise

$$\alpha\ \underline{Q}C\ \omega$$
$$(\alpha\ \underline{Q}G\ \omega)\times(\alpha\times\alpha\ \underline{Q}G\ \omega)\ \underline{Q}F\ G\ \omega$$

The proofs of the difference-quotient rules follow from the fact that if K is a dyadic function for which

$$F\ \omega+\alpha \qquad [1.3.1]$$
$$(F\ \omega)+\alpha\times\alpha\ K\ \omega$$

when $\alpha\neq 0$, then K must be identical to $\underline{Q}F$ for non-zero left arguments. This is an obvious fact, since Identity 1.3.1 is equivalent to

$$((F\ \omega+\alpha)-F\ \omega)\div\alpha$$
$$\alpha\ K\ \omega$$

when $0\neq\alpha$. In particular:

$$F\ \omega+\alpha \qquad [1.3.2]$$
$$(F\ \omega)+\alpha\times\alpha\ \underline{Q}F\ \omega$$

when $\alpha \neq 0$. We may say that the difference-quotient function $\underline{Q}F$ is uniquely determined by Identity 1.3.1. Note that the zero left argument has been excluded, since the values $0\ \underline{D}QF\ X$ are the same as $0 \div 0$, and are therefore indeterminate. Identities 1.3.1 and 1.3.2 are called <u>addition formulas</u> for F because they express relations between $F\ \omega+\alpha$ and $F\ \omega$. There can be many forms of identities relating $F\ \omega+\alpha$ and $F\ \omega$; Identities 1.3.1 and 1.3.2 are said to be addition formulas of <u>normal form</u>.

The Plus Rule for Difference-Quotients may now be proved by considering the following sequence of identities when $\alpha \neq 0$:

$$P\ \omega+\alpha$$
$$(F\ \omega+\alpha)+G\ \omega+\alpha$$
$$((F\ \omega)+\alpha\times\alpha\ \underline{Q}F\ \omega)+(G\ \omega)+\alpha\times\alpha\ \underline{Q}G\ \omega$$
$$((F\ \omega)+G\ \omega)+\alpha\times(\alpha\ \underline{Q}F\ \omega)+\alpha\ \underline{Q}G\ \omega$$
$$(P\ \omega)+\alpha\times(\alpha\ \underline{Q}F\ \omega)+\alpha\ \underline{Q}G\ \omega$$

Evidently, if $K{:}(\alpha\ \underline{Q}F\ \omega)+\alpha\ \underline{Q}G\ \omega$ then the last expression in this sequence is equivalent to $(P\ \omega)+\alpha\times\alpha\ K\ \omega$. Consequently $P\ \omega+\alpha$ is equivalent to $(P\ \omega)+\alpha\times\alpha\ K\ \omega$ when $\alpha \neq 0$, and therefore K must be equivalent to the difference-quotient of $(F\ \omega)+G\ \omega$.

The proof of the Times Rule for Difference-Quotients is similar. Namely, if $T\ \omega$ equals $(F\ \omega)\times G\ \omega$ then we have the following sequence of identities when $\alpha \neq 0$:

$$T\ \omega+\alpha$$
$$(F\ \omega+\alpha)\times G\ \omega+\alpha$$
$$((F\ \omega)+\alpha\times\alpha\ \underline{Q}F\ \omega)\times(G\ \omega)+\alpha\times\alpha\ \underline{Q}G\ \omega$$
$$((F\ \omega)\times G\ \omega)+\alpha\times((\alpha\ \underline{Q}F\ \omega)\times G\ \omega)+((F\ \omega)\times\alpha\ \underline{Q}G\ \omega)+$$
$$\alpha\times(\alpha\ \underline{Q}F\ \omega)\times\alpha\ \underline{Q}G\ \omega$$

Thus if

$$K{:}((\alpha\ \underline{Q}F\ \omega)\times G\ \omega)+((F\ \omega)\times\alpha\ \underline{Q}G\ \omega)+\alpha\times(\alpha\ \underline{Q}F\ \omega)\times\alpha\ \underline{Q}G\ \omega$$

then the last expression in this sequence is equivalent to $(T\ \omega)+\alpha\times\alpha\ K\ \omega$. Consequently $T\ \omega+\alpha$ is equivalent to $(T\ \omega)+\ \alpha\times\alpha\ K\ \omega$, and therefore K must be equivalent to the difference-quotient of $(F\ \omega)\times G\ \omega$.

Proceeding to the Reciprocal Rule for Difference-Quotients, we have the following sequence when $\alpha \neq 0$:

$$R\ \omega+\alpha$$
$$\div F\ \omega+\alpha$$
$$\div(F\ \omega)+\alpha\times\alpha\ \underline{Q}F\ \omega$$
$$\div(F\ \omega)\times 1+\alpha\times(\alpha\ \underline{Q}F\ \omega)\div F\ \omega$$
$$(\div F\ \omega)\times\div 1+\alpha\times(\alpha\ \underline{Q}F\ \omega)\div F\ \omega$$
$$(R\ \omega)\times\div 1+\alpha\times(\alpha\ \underline{Q}F\ \omega)\div F\ \omega$$

Using the fact that ÷1+ω and 1-ω÷1+ω are equivalent, and substituting α×(α Q̲F ω)÷F ω for ω in these expressions, we continue the above sequence as follows:

```
(R ω)×1-(α×(α Q̲F ω)÷F ω)÷1+α×(α Q̲F ω)÷F ω
(R ω)+α×-((R ω)×(α Q̲F ω)÷F ω)÷1+α×(α Q̲F ω)÷F ω
(R ω)+α×-(α Q̲F ω)÷((F ω)*2)×1+α×(α Q̲F ω)÷F ω
```

Arguing as before, if

```
K:-(α Q̲F ω)÷((F ω)*2)×1+α×(α Q̲F ω)÷F ω
```

then the above sequence shows that H ω+α is equivalent to (F ω)+α×α K ω when α≠0. Consequently K is equivalent to the difference-quotient of ÷F ω.

Finally, we come to the Composite Rule for Difference-Quotients. If S Q̲G X is 0 then G X+S equals G X. Consequently F G X+S equals F G X and S C̲C X is 0. Otherwise, we have the following sequence for α≠0:

```
C ω+α
F G ω+α
F (G ω)+α×α Q̲G ω
```

Substituting X for G ω and S for α×α Q̲G ω in the last expression permits this sequence to be continued as follows:

```
F X+S
(F X)+S×S Q̲F X
```

Replacing G ω for X and α×α Q̲G ω for S in the latter expression, we continue as follows:

```
(F G ω)+α×(α Q̲G ω)×(α×α Q̲G ω) Q̲F G ω
```

Therefore K:(α Q̲G ω)×(α×α Q̲G ω) Q̲F G ω is equivalent to the difference-quotient of F G ω. ⌹37

We turn now to the equivalent expressions for the difference-quotients of the functions ω*N provided by the Binomial Theorem, where N is a non-negative integer. For example, if N is 2 we have the following sequence:

```
(ω+α)*2
+/(ω*0 1 2)×1 2 1×α*2 1 0
(ω*2)++/¯1↓(ω*0 1 2)×1 2 1×α*2 1 0
(ω*2)+α×+/(ω*0 1)×1 2×α*1 0
```

and if N is 3:

```
(ω+α)*3
+/(ω*0 1 2 3)×1 3 3 1×α*3 2 1 0
(ω*3)++/¯1↓(ω*0 1 2 3)×1 3 3 1×α*3 2 1 0
(ω*3)+α×+/(ω*0 1 2)×1 3 3×α*2 1 0
```

Therefore the difference-quotients of `ω*2` and `ω*3` are equivalent to `+/(ω*0 1)×1 2×α*1 0` and `+/(ω*0 1 2)×1 3 3×α*2 1 0` respectively. The vectors `1 2 1` and `1 3 3 1` in the above expressions are called vectors of <u>binomial coefficients</u> and are produced by the function

```
BIN:(0,BIN ω-1)+(BIN ω-1),0 : 0=ω : 1⍴1
```

For example:

```
      BIN 2
1 2 1
      BIN 3
1 3 3 1
```

<u>The Binomial Theorem</u>. For a non-negative integer `N`:

```
      (ω+α)*N                                    [1.3.3]
      +/(ω*⍳N+1)×(BIN N)×α*⌽⍳N+1
      (ω*N)+α×+/(ω*⍳N)×(¯1↓BIN N)×α*⌽⍳N          ⊞39
```

Using the Binomial Theorem as a guide, we define the functions `P:ω*N` and `CP:+/(ω*⍳N)×(¯1↓BIN N)×α*⌽⍳N`. We then have the following identity for `P`:

```
      P ω+α
      (P ω)+α×α CP ω
```

Comparing this identity with 1.3.1, we see that `CP` must be equivalent to the difference-quotient function `QP:((P ω+α)-P ω)÷α` when `α≠0`. Moreover, `CP` is also defined when `α=0`, whereas `0 QP ω` is the same as `0÷0`, and is therefore indeterminate. In fact, for the function `0 CP ω` we have the identity:

```
      0 CP ω                                     [1.3.4]
      N×ω*N-0≠N
```

To see that this is true for a positive integer `N`, we begin with the identity:

```
      0 CP ω
      +/(ω*⍳N)×(¯1↓BIN N)×0*⌽⍳N
```

Since

```
      0*⌽⍳N
      (-N)↑1
```

all elements of the vector `(ω*⍳N)×(¯1↓BIN N)×0*⌽⍳N` are zero except the last. The last element of `ω*⍳N` is `ω*N-1`, and the last element of `¯1↓BIN N` is `(BIN N)[N-1]`. Therefore

```
      0 CP ω
      +/(ω*⍳N)×(¯1↓BIN N)×0*⌽⍳N
      (BIN N)[N-1]×ω*N-1
```

It is not difficult to check that (*BIN N*)[*N*-1] is *N*, and this completes the proof for a positive integer *N*. The proof when *N* is 0 is left to the exercises.

The function *C̲P* is called the <u>cline</u> or <u>cline</u> <u>function</u> of *P*. ⌹44

1.4 CLINE FUNCTIONS

To each function *F* there is an associated difference-quotient *Q̲F*:((*F* ω+α)-*F* ω)÷α, and if *P*:ω⋆*N* for a non-negative integer *N*, then there is also an associated cline function *C̲P*:+/(ω⋆⍳*N*)×(¯1↓*BIN N*)×α⋆⌽⍳*N*, which agrees with *Q̲P* and is also defined for zero left argument. We will now present a general definition of cline functions that is not restricted to the functions ω⋆*N*. We begin by examining a geometric interpretation of the values of the cline function *C̲P*, from which this function derives its name.

The values of the cline function *C̲P* coincide with the values of the difference-quotient *Q̲P*:((*P* ω+α)-*P* ω)÷α for non-zero left arguments. Consequently, the value *S C̲P X* is the slope of the secant line joining the points *X*,*P X* and (*X*+*S*),*P X*+*S* on the graph of *P* for non-zero scalars *S*. As the scalar *S* is chosen closer and closer to zero, the point (*X*+*S*),*P X*+*S* moves along the graph of *P* towards the point *X*,*P X*, and the secant line passing through the points *X*,*P X* and (*X*+*S*),*P X*+*S* rotates towards the line tangent to the graph of *P* at the point *X*,*P X*, as illustrated in Figure 1.4.1. Since the values *S C̲P X* are the slopes of the secant

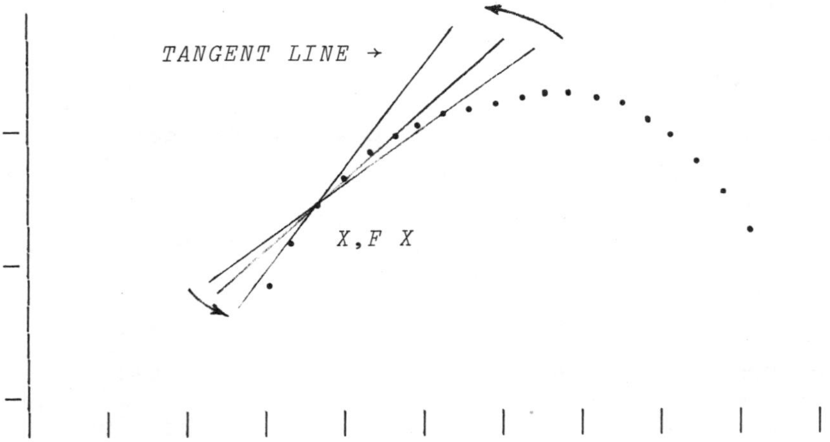

SECANT LINES ROTATE TOWARDS TANGENT LINE
FIGURE 1.4.1

lines passing through the points *X*,*P X* and (*X*+*S*),*P X*+*S*, it is reasonable to conclude that the values *S C̲P X* approach the slope of the line tangent to the graph of *F* at the point *X*,*P X* as the scalar *S* is chosen closer and closer to zero. It would then be true that the value 0 *C̲P X* would be the slope of this tangent line, because the values *S C̲P X* also approach the value 0 *C̲P X* as *S* is chosen closer and closer to zero.

To summarize, for a non-zero scalar *S* the value *S C̲P X* is the slope of the secant line joining the points *X*,*P X* and (*X*+*S*),*P X*+*S* on the graph of *P*, while the value 0 *C̲P X* is the slope of the line tangent to the graph of *P* at the point *X*,*P X*.

More generally, the cline *C̲F* of a function *F* is defined so that the geometric interpretations of its values are the same as those for the cline of ω⋆*N*. Thus:

The Definition of Cline Functions. The cline function *C̲F* of a function *F* is defined to be equivalent to the difference-quotient *Q̲F* for non-zero left arguments, while the value 0 *C̲F X* is defined to be the value approached by the values *S Q̲F X* as *S* is chosen closer and closer to zero. The value 0 *C̲F X* is not defined if the values *S C̲F X* do not approach a unique value as *S* is chosen closer and closer to zero.

In particular, since the values *S Q̲F X* are slopes of secant lines that rotate towards the line tangent to the graph of *F* at the point *X*,*F X* as *S* is chosen closer and closer to zero, the value 0 *C̲F X* is the slope of that tangent line.

The general definition of cline functions is partially implicit, in that no explicit rule for producing the values 0 *C̲F X* is given. However, the rules for difference-quotients can be used to derive rules for cline functions, so that the cline of a function defined by a complicated expression can be analyzed in terms of the clines of its constituents. In particular, since we have explicit definitions of the clines of the functions ω⋆*N*, we can produce explicit definitions of the clines of combinations of the functions ω⋆*N*.

For example, consider the function *H*:(ω⋆2)+ω⋆3, which is the sum of the functions *F*:ω⋆2 and *G*:ω⋆3. We will show in a moment that α *C̲H* ω is equivalent to (α *C̲F* ω)+α *C̲G* ω. Thus, since we know explicit definitions of *C̲F* and *C̲G*, we can produce an explicit definition of *C̲H*, namely, *C̲H*:(α *C̲F* ω)+α *C̲G* ω, or

```
C̲H:(+/(ω⋆0 1)×1 2×α⋆1 0)++/(ω⋆0 1 2)×1 3 3×α⋆2 1 0
```

Plus Rule for Cline Functions. If *P ω* is equivalent to *(F ω)+G ω*, then

```
α C̲P ω
(α C̲F ω)+α C̲G ω
```

To prove the Plus Rule for Cline Functions, we use the Plus Rule for Difference-Quotients, which states that *α Q̲H ω* is equivalent to *(α Q̲F ω)+α Q̲G ω*. Therefore *α C̲H ω* is equivalent to *(α C̲F ω)+α C̲G ω* when *α≠0*. As the scalar *S* is chosen closer and closer to zero, the values *S C̲H X* approach *0 C̲H X*; these values also are equal to *(S C̲F X)+S C̲G X*, which in turn approach *(0 C̲F X)+0 C̲G X*. Thus *0 C̲H X* is equal to *(0 C̲F X)+0 C̲G X*. In other words, *α C̲H ω* is also equivalent to *(α C̲F ω)+α C̲G ω* when *α=0*, and this completes the proof of the plus rule.

Thus the cline of *P* is equivalent to the sum of the clines of *F* and *G*. In particular, if explicit definitions of *C̲F* and *C̲G* are known, their sum is then an explicit definition of *C̲P*. ⍠47

The arguments used to extend the Plus Rule for Difference-Quotients to the Plus Rule for Cline Functions carry over to the extension of the other difference-quotient rules to corresponding rules for cline functions. The resulting rules for cline functions are as follows:

Times Rule for Cline Functions. If *T ω* is equivalent to *(F ω)×G ω*, then

```
α C̲T ω
((α C̲F ω)×G ω)+((F ω)×α C̲G ω)+α×(α C̲F ω)×α C̲G ω
```

Reciprocal Rule for Cline Functions. If *R ω* is equivalent to *÷F ω*, then

```
α C̲R ω
-(α C̲F ω)÷((F ω)*2)×1+α×(α C̲F ω)÷F ω
```

Composite Rule for Cline Functions. If *C ω* is equivalent to *F G ω*, then

```
α C̲C ω
(α C̲G ω)×(α×α C̲G ω) C̲F G ω
```
⍠48

The function *0 C̲F ω* is defined only for arguments for which the graph of *F* is a "smooth" curve. That is, *0 C̲F X* is not defined if there is a break or sharp peak in the graph of *F* at *X*, since in that event the values *S Q̲F X* do not approach a unique value as *S* is chosen closer and closer to zero.

For example, the graph of *F*:×ω-1 in Figure 1.4.2 has a break at 1 and the graph of *G*:2+|ω-1 in Figure 1.4.3 has a sharp peak at 1. Moreover, we have the following identities when α≠0:

α $\underline{C}F$ 1
(×α)÷α

α $\underline{C}G$ 1
×α

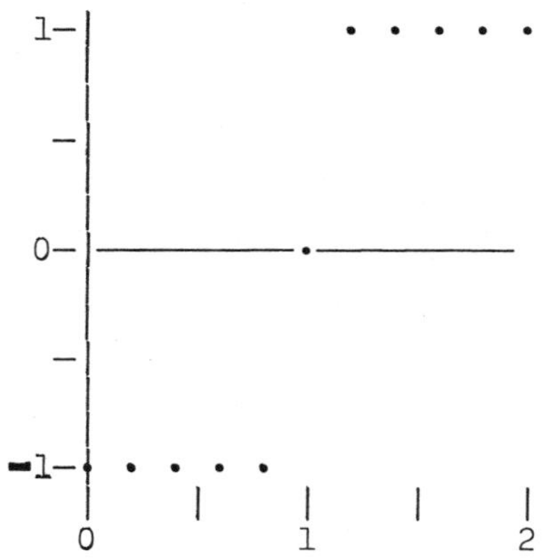

GRAPH OF ×ω-1
FIGURE 1.4.2

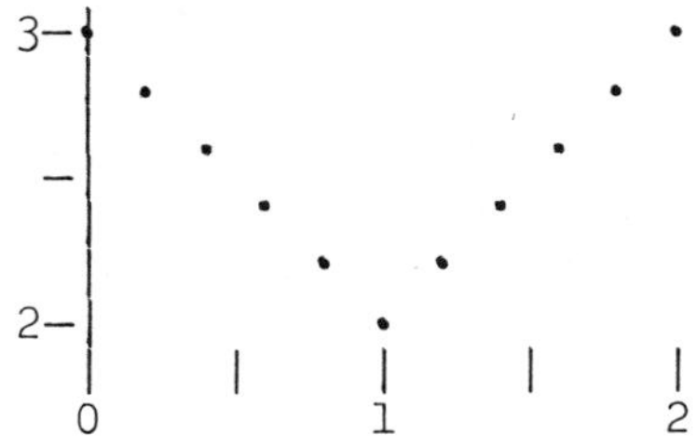

GRAPH OF 2+|ω-1
FIGURE 1.4.3

Thus as *S* is assigned values closer and closer to zero, the values *S* $\underline{C}F$ 1 grow indefinitely large in magnitude, and therefore do not approach a unique value. Also, the values *S* $\underline{C}G$ 1 are either 1 or ¯1 according to whether *S* is positive or negative, and therefore also do not approach a unique value. Thus neither 0 $\underline{C}F$ 1 nor 0 $\underline{C}G$ 1 is defined.

1.5 DERIVATIVES

We have seen in Section 1.2 that a sketch of the graph of a function *F* can be produced from segments of secant lines passing through pairs of neighboring points on the graph. Graph sketches can also be produced from segments of tangent lines. For example, consider again the function *F*:¯2+(¯1×ω)+ω*2 and the vectors *X*←¯2 1 3.5 and *F X* of Section 1.2. The line segments in Figure 1.5.1 are segments of the lines tangent to the graph of *F* at the coordinate points *X*[*I*],*F X*[*I*] when *I* is 0, 1 and 2. The graph in Figure 1.5.1 is produced by drawing a smooth curve tangent to each line segment at the appropriate point.

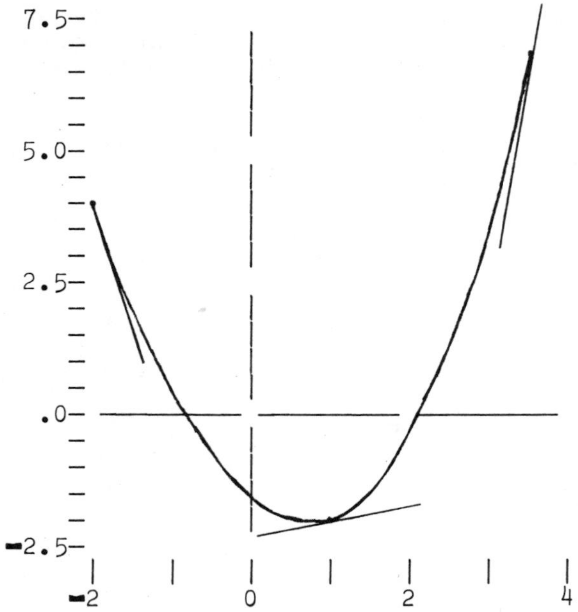

GRAPH PRODUCED FROM TANGENT LINES
FIGURE 1.5.1

At first thought it may appear that we have reversed the natural order of things, in that sketches of tangent lines should be produced from sketches of graphs and not vice-versa. However, we can easily sketch tangent lines if we know their slopes, and the slopes can be computed from cline functions. Precisely, the slope of the line tangent to the graph of the function F at the point $X,F\ X$ is $0\ \underline{C}F\ X$.

We will examine the use of tangent lines in graph sketching in the exercises. The remainder of this section is devoted to derivatives, which are functions whose values are slopes of tangent lines.

<u>The Definition of the Derivative</u>. The <u>derivative</u> of a function F, which will be denoted by $\underline{D}F$, is defined in terms of the cline of F as follows:

$$\underline{D}F{:}0\ \underline{C}F\ \omega$$

In particular, the value $\underline{D}F\ X$ is the slope of the line tangent to the graph of F at the point $X,F\ X$. ⊞56

If the function G is a weighted sum of the functions $\alpha\star N$ for non-negative integers N, then its cline function $\underline{C}G$ can be produced by the rules of the preceding section and in turn its derivative $\underline{D}G$ can be defined as $\underline{D}G{:}0\ \underline{C}G\ \omega$. We will now examine a more direct route to an explicit definition of $\underline{D}G$. Namely, the rules for cline functions lead to rules for derivatives, which in turn lead to an explicit definition of $\underline{D}G$.

<u>The Derivative of the Function</u> $\omega\star N$. If N is a positive integer, then the derivative of the function $P{:}\omega\star N$ is $\underline{D}P{:}N\times\omega\star N\text{-}0\neq N$. For example, the derivative of $\omega\star 2$ is $2\times\omega\star 1$ and the derivative of $\omega\star 3$ is $3\times\omega\star 2$. The foregoing fact follows directly from the definition of derivatives and Identity 1.3.4. ⊡59

<u>The Plus Rule for Derivatives.</u> If $P\ \omega$ is equivalent to $(F\ \omega)+G\ \omega$, then

$\underline{D}P\ \omega$
$(\underline{D}F\ \omega)+\underline{D}G\ \omega$

The proof of this rule follows from the Plus Rule for Cline Functions:

$\underline{D}P\ \omega$
$0\ \underline{C}P\ \omega$
$(0\ \underline{C}F\ \omega)+0\ \underline{C}G\ \omega$ — Plus Rule for Clines
$(\underline{D}F\ \omega)+\underline{D}G\ \omega$

For example, if $P{:}(\omega\star 2)+\omega\star 5$ then $\underline{D}P{:}(2\times\omega)+5\times\omega\star 4$. ⊡60

<u>The Times Rule for Derivatives.</u> If $T\ \omega$ is equivalent to $(F\ \omega)\times G\ \omega$, then

$\underline{D}T\ \omega$
$((F\ \omega)\times\underline{D}G\ \omega)+(\underline{D}F\ \omega)\times G\ \omega$

As with the plus rule, the times rule follows from the Times Rule for Cline Functions:

$\underline{D}T\ \omega$
$0\ \underline{C}T\ \omega$
$((F\ \omega)\times 0\ \underline{C}G\ \omega)+((0\ \underline{C}F\ \omega)\times G\ \omega)+0\times(0\ \underline{C}F\ \omega)\times 0\ \underline{C}G\ \omega$
$((F\ \omega)\times\underline{D}G\ \omega)+(\underline{D}F\ \omega)\times G\ \omega$

For example, suppose that $T{:}(\omega\star 3)\times\omega+\omega\star 2$. If we define $F{:}\omega\star 3$ and $G{:}\omega+\omega\star 2$, then $T\ \omega$ is identical to $(F\ \omega)\times G\ \omega$. The derivatives of F and G are $\underline{D}F{:}3\times\omega\star 2$ and $\underline{D}G{:}1+2\times\omega$ respectively. Therefore $\underline{D}T{:}((F\ \omega)\times\underline{D}G\ \omega)+(\underline{D}F\ \omega)\times G\ \omega$ or, equivalently, $\underline{D}T{:}((\omega\star 3)\times 1+2\times\omega)+(3\times\omega\star 2)\times\omega+\omega\star 2$.

An important special case of the times rule occurs when $F{:}B+0\times\omega$ for a scalar B. According to Exercise 1.50, $\underline{C}F{:}0\times\alpha+\omega$ and therefore $\underline{D}F{:}0\times\omega$. Thus if $H{:}B\times G\ \omega$, then $\underline{D}H{:}(B\times\underline{D}G\ \omega)+0\times G\ \omega$, or equivalently $\underline{D}H{:}B\times G\ \omega$. In words, the derivative of a constant times $G\ \omega$ is that constant times $\underline{D}G\ \omega$.

For example, the derivative of *H*:5×ω⋆3 is D̲*H*:5×3×ω⋆2. Also, combining this rule with the sum rule, we see that the derivative of *H*:3+(¯4×ω)+ω⋆2 is D̲*H*:0+(¯4×1×ω⋆0)+2×ω⋆1, or equivalently D̲*H*:¯4+2×ω. ⌹61

<u>The Reciprocal Rule for Derivatives</u>. If *R* ω is equivalent to ÷*F* ω, then

```
D̲R ω
-(D̲F ω)÷(F ω)⋆2
```

since

```
D̲R ω
0 C̲R ω
-(0 C̲F ω)÷((F ω)⋆2)×1+0×(0 C̲F ω)÷F ω
-(D̲F ω)÷(F ω)⋆2
```

For example, if *R*:÷1+ω⋆2 and we define *F*:1+ω⋆2, then *R* ω is equivalent to ÷*F* ω. Therefore D̲*R*:-(2×ω)÷(*F* ω)⋆2, or equivalently D̲*H*:-(2×ω)÷(1+ω⋆2)⋆2.

Also, it follows from the Reciprocal Rule for Derivatives that if *H*:ω⋆*N* for a negative integer *N*, then D̲*H*:*N*×ω⋆*N*-1. The proof will be developed in the exercises. Thus the derivative rule for the functions ω⋆*N* holds more generally than was originally stated. ⌹65

<u>The Composition Rule for Derivatives</u>. If *C* ω is equivalent to *F G* ω, then

```
D̲C ω
(D̲F G ω)×D̲G ω
```

In words, the function value D̲*H X* is identical to D̲*G X* times the value of D̲*F* at *G X*. The proof is as follows:

```
D̲H ω
0 C̲H ω
(0 C̲G ω)×(0×0 C̲G ω) C̲F̲ G ω
(0 C̲G ω)×0 C̲F G ω
(D̲G ω)×D̲F G ω
```

For example, suppose that *C*:(3+ω⋆2)⋆4. If we define *F*:ω⋆4 and *G*:3+ω⋆2 then *C* ω is identical to *F G* ω. Evidently D̲*F*:4×ω⋆3, and D̲*G*:2×ω, and D̲*F G* ω is identical to 4×(3+ω⋆2)⋆3. Consequently D̲*C*:(2×ω)×4×(3+ω⋆2)⋆3.

The Composition Rule for Derivatives is also called the <u>chain rule</u>, since it produces derivatives of functions defined by "chaining" other functions together. ⌹70

1.6 THE DERIVATIVE AS RATE OF CHANGE

For a function *F* and arguments *X* and *S* for which *S*≠0, the value (*F X*+*S*)-*F X* represents the change in values of *F* as the argument *X* is replaced by *X+S*. From this viewpoint the function value *S QF X*, where *QF* is the difference-quotient function *QF*:((*F* ω+α)-*F* ω)÷α, can be thought of as an average rate of change in the values of *F* for arguments between *X* and *X+S*. If the value of *S* is small then the function value *S QF X* is approximately *DF X*, where *DF* is the derivative of *F*, and therefore *DF X* can be thought of as the rate of change in the values of *F* at *X*.

For example, consider the function *F*:ω*2, for which *DF*:2×ω. Since *DF* 2 is 4, the rate of change in the values of *F* at 2 is 4. An examination of the graph of *F* in Figure 1.6.1 shows that near the coordinate point 2 4 the values of *F* are changing at a rate roughly 4 times the rate at which the argument changes. Also, since *DF* ¯1 is ¯2, the rate of change in the values of *F* at ¯1 is ¯2. As before, an examination of the graph of *F* shows that near the coordinate point ¯1 1 the values of *F* are changing at a rate roughly ¯2 times the rate at which the argument changes.

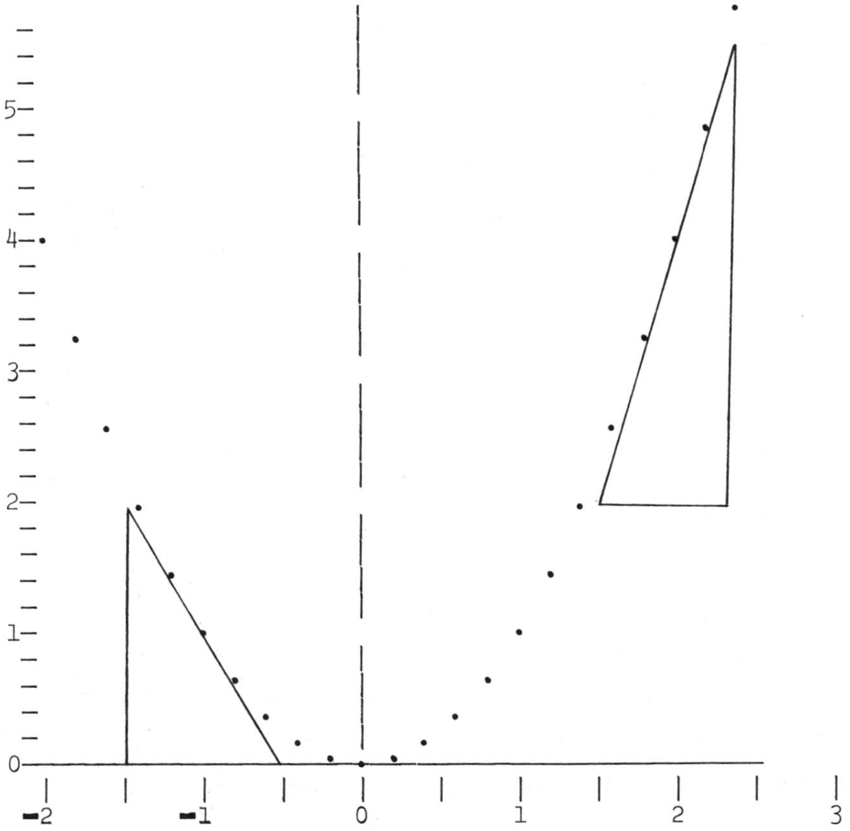

CHANGE IN VALUES VS. CHANGE IN ARGUMENTS
FIGURE 1.6.1

The sign of the value *D̲F X* indicates whether or not the function values are increasing or decreasing as the arguments near *X* increase. For example, *D̲F* 2 is positive and the values of *F* increase as the arguments near 2 increase. Also, *D̲F* ¯1 is negative and the values of *F* decrease as arguments near ¯1 increase. ⍠72

There is special meaning to the derivative as rate of change when the function represents motion. Suppose that the function *POS* represents the position of a moving object; that is, the function value *POS T* is the position of the object at time *T*. Then the value *S Q̲POS T* is an average velocity of the object as it moves from the position *POS T* to the position *POS T+S*. If the function *VEL* is the velocity function of the object, in that the function value *VEL T* is the velocity at time *T*, then *VEL T* is approximately *S Q̲POS T* for small values of *S*. Moreover, the value *D̲POS T* of the derivative of *POS* is also approximately *S Q̲POS T* for small values of *S*. Thus we should suspect that *D̲POS* is *VEL*. That is, the derivative of the position function is identical to the velocity function.

Continuing this functional description of motion, the value *S Q̲VEL T* is an average acceleration of the object as it moves from position *POS T* to position *POS T+S*. If the function *ACC* is the acceleration function of the object, that is, if the function value *ACC T* is the acceleration at time *T*, then *ACC T* is approximately *S Q̲VEL T* for small values of *S*. Moreover, the value *D̲VEL T* of the derivative of *VEL* is also approximately *S Q̲VEL T* for small values of *S*. Thus the derivative of the velocity function is the acceleration function.

Since the derivative of *POS* is *VEL* and the derivative of *VEL* is *ACC*, the derivative of the derivative of *POS* is *ACC*. The derivative of the derivative of *POS* is called the <u>second derivative</u> of *POS* and will be denoted by *D̲D̲POS*. Consequently *D̲D̲POS ω* is *ACC ω*. If, as is true in many mechanical systems, acceleration is also a function of position then there is a function *ACP* for which *ACC ω* is *ACP POS ω*. Thus we have the identity:

D̲D̲POS ω [1.6.2]
ACP POS ω

This identity is called a <u>differential equation</u> because it expresses a relation between a function (in this case *POS*) and its derivatives (in this case the second derivative). The function *POS* is called a <u>solution</u> of the differential equation. Differential equations can be thought of as implicit definitions of their solutions. In Chapter 6 we will examine procedures for producing explicit, but approximate definitions of the solutions, which are some of the ways the calculus provides for constructing new functions.

1.7 INTEGRALS

The process of producing the derivative *D̲F* of a function *F* is called differentiation. Roughly speaking, this process can be reversed so that *F* can be produced from *D̲F*. The reverse process is called integration and in this regard *F* is called an integral or anti-derivative of *D̲F*. That is, if the derivative of the function *H* is the function *G*, then *H* is called an integral or anti-derivative of *G*.

It is not difficult to produce integrals of weighted sums of the functions ω*N for integers *N*. For example, if G:ω*3 then H:(÷4)×ω*4 is an integral of *G*; if G:5×ω*7 then H:(5÷8)×ω*8 is an integral of *G*; and if G:3+(¯4×ω)+ω*4 then H:(3×ω)+((¯4÷2)×ω*2)+(÷5)×ω*5 is an integral of *G*. In each case we show that *H* is an integral of *G* by producing the derivative *D̲H* and showing that *D̲H* is equivalent to *G*. For instance, in the first example the derivative of H:(÷4)×ω*4 is D̲H:4×(÷4)÷ω*3 or equivalently D̲H:ω*3, which is also *G*. Thus *H* is an integral of *G*.

An Integral of ω*N. If *N* is an integer not equal to ¯1 then Q:(÷N+1)×ω*N+1 is an integral of P:ω*N.

This statement follows from the fact that the derivative of *Q* is D̲Q:(÷N+1)×(N+1)×ω*N, or equivalently D̲Q:ω*N. For example, Q:(÷5)×ω*5 is an integral of P:ω*4. Note that this rule does not apply if *N* is ¯1.

The Plus Rule For Integrals. If *H1* and *H2* are integrals of *G1* and *G2* respectively then H:(H1 ω)+H2 ω is an integral of G:(G1 ω)+G2 ω.

This rule follows from the fact that derivative of *H* is the sum of the derivatives of *H1* and *H2*, which is the sum of *G1* and *G2*, or equivalently *G*. For example, H:((÷2)×ω*2)+(÷4)×ω*4 is an integral of G:ω+ω*3.

The Scale Rule for Integrals. The scale rule is concerned with integrals of the function G SC ω, where the function SC:A+B×ω is called a scale function. Namely, if *H* is an integral of *G* then

(÷B)×H SC ω

is an integral of G SC ω.

For example, the function (÷5)×(÷3)×(¯4+5×ω)*3 is an integral of (¯4+5×ω)*2. The scale rule is demonstrated by producing the derivative of (÷B)×H SC ω and comparing it to G SC ω.

Other integral rules will be presented in Chapter 8. The reason for deferring the discussion of these rules is that procedures for producing explicit definitions of integrals are not as straightforward as those for derivatives. For instance, we know a rule for producing the derivatives of all functions of the form ω*N, where N is an integer. This is not true for integrals, since we do not yet know an explicit definition of an integral of ω*¯1. ⌹74

Evidently if H is an integral of G and C is a constant then K:C+H ω is also an integral of G. For example, H:ω*3 is an integral of G:3×ω*2, as are K:5+ω*3 and L:¯14+ω*3, and so on. Thus while a function has only one derivative, it has many integrals. Moreover, any integral of a function whose domain is an interval is obtained by adding a constant to any one particular integral. This fact, whose proof is discussed in a later exercise, enables us to produce an expression for defining a function F from an expression for its derivative and one value of F. For example, suppose that D̲F:3+ω and F 0 is 5. The function H:(3×ω)+(÷2)×ω*2 is an integral of D̲F. According to the fact stated above, since F is also an integral of D̲F there must be a constant C for which F ω is identical to C+H ω. In particular, C+H 0, which is C+0 or C, must be identical to F 0, which is 5. That is, C is 5 and consequently F:5+H ω, or equivalently F:5+(3×ω)+(÷2)×ω*2.

As another example, suppose that D̲F:4×ω*3 and F 2 is 5. Then H:ω*4 is an integral of D̲F, as is C+H ω for a constant C. If C is the constant for which C+H ω is identical to F ω, then in particular C+H 2, which is C+16, is identical F 2, which is 5. Thus C is ¯11, and F:¯11+ω*4. ⌹75

In general, if H is an integral of F then

B+(H ω)-H A

is an integral of F whose value at A is B.

As a final example, consider the motion of a ball thrown straight upward near the earth. The acceleration of the ball is ¯32 feet per second per second; acceleration is negative because it is directed towards the earth, and it is usual to take the sign of the direction away from the earth to be positive. Thus ACC:¯32+0×ω. Similarly, a positive value for velocity indicates that the ball is rising away from the earth, while a negative value indicates that the ball is falling towards the earth. The meaning of positive and negative values of the altitude of the ball should be obvious.

Since ACC is the derivative of the velocity function VEL, VEL is an integral of ACC and so there is a constant V for which VEL:V+¯32×ω. Note that V is VEL 0, the <u>initial velocity</u> of the ball. Also, since VEL is the derivative of POS, where the values of POS are the ball's altitude, POS is an integral of *VEL* and so there is a constant *A* such that

    POS:A+(V×ω)+¯32×(÷2)×ω*2

or equivalently

    POS:A+(V×ω)+¯16×ω*2

Note that *A* is *POS* 0, the initial position of the ball. Thus if a ball is thrown into the air from a height of 100 feet with an initial velocity of 10 feet per second, its altitude *T* seconds after it was released is *POS T*, where *POS*:100+(10×ω)+¯16×ω*2.

More generally, for constants *T*0, *V*0 and *A*0 the velocity function *VEL* and the position function *POS* can be defined as

    VEL:V0+¯32×ω-T0

and

    POS:A0+(V0×ω-T0)+¯16×(ω-T0)*2

These definitions permit us to easily define velocity and position functions when their values are known for an argument other than zero. For example, suppose that the velocity of a ball is ¯17 feet per second 3 seconds after it is thrown, and at the same instant its altitude is 50 feet. Then its altitude *T* seconds after it was released is *POS T*, where *POS*:50+(¯17×ω-3)+¯16×(ω-3)*2. ⌹77

Integrals will sometimes be named in a way similar to that for naming derivatives, using the letter *I* as the prefix. For example, *IF*:ω*2 is an integral of *F*:2×ω. The underscored letter *I̲* will not be used as the prefix because functions have many integrals and underscored letters will be reserved for prefixes of unique associated functions.

We have now introduced the idea of the derivative, produced derivatives of polynomials, that is, weighted sums of the functions ω*N for non-negative integers *N*, and applied derivatives to graph sketching and the description of the motion of a projectile. We have still to look more closely at the derivative as a tool for analyzing and constructing functions. Before we do this we will study linear functions (Chapter 2) and the algebra of polynomials (Chapter 3), returning to applications of the derivative in Chapter 4.

2
The Matrix Product

2.1 THE DEFINITION OF THE MATRIX PRODUCT

For <u>conformable</u> matrices *A* and *B* (for which the number of columns of *A* equals the number of rows of *B*), the <u>matrix product</u> or +.× <u>inner product</u> of *A* and *B* is the matrix denoted by *A*+.×*B* and defined by the identity

```
      (A+.×B)[I;J]
      +/A[I;]×B[;J]
```

In words, the element at row *I*, column *J* of *A*+.×*B* is identical to the plus reduction of row *I* of *A* times column *J* of *B*. For example:

```
      A←2 3ρ2 8 5 6 3 1
      A
2  8  5
6  3  1
      B←3 5ρ7 7 10 4 6 9 1 1 6 7 1 4 1 5 7
      B
7  7  10  4  6
9  1   1  6  7
1  4   1  5  7

      A+.×B
91  42  33  81  103
70  49  64  47   64

      I←1
      J←0
      (A+.×B)[I;J]              +/A[I;]×B[;J]
70                        70                          ⌹1
```

The number of rows of *A*+.×*B* is identical to the number of rows of the left argument *A* and the number of columns of *A*+.×*B* is identical to the number of columns of the right argument *B*. Formally:

```
      ρA+.×B                                    [2.1.1]
      (¯1↓ρA),1↓ρB                                   ⌹3
```

To continue the preceding example, ¯1↓ρ*A* is 2, 1↓ρ*B* is 5, and ρ*A*+.×*B* is 2 5.

The basic facts concerning the matrix product can be kept in mind with the diagram of Figure 2.1.2. The block labelled *L* represents the left argument of a matrix product and the block labelled *R* the right argument. The block labelled *SQUARE* is a square block to indicate that the number of columns of *L* is identical to the number of rows of *R*. The name *RV* refers to a row of *L* and *CV* refers to a column of *R*. The diagram indicates that the row *I*, column *J* element of *L*+.×*R* is produced from row *I* of *L* and column *J* of *R*. The size of the block labelled *L*+.×*R* indicates that the number of rows of the matrix product *L*+.×*R* is identical to the number of rows of the left argument *L* and the number of columns of this matrix product is identical to the number of columns of the right argument *R*. ⌹7

```
 ---------   --------------
|  L       || L+.×R   |      |
|____RV____||_________|______|
|          ||         |      |
 ---------   --------------
 ---------   --------------
|          ||     R   |      |
|          ||         |      |
|  SQUARE  ||         |CV    |
|          ||         |      |
|          ||         |      |
 ---------   --------------
```

THE +.× INNER PRODUCT
FIGURE 2.1.2

The importance of the +.× inner product is due to the following facts:

> Many functions can be represented in the form *A*+.×*ω*. These functions, known as linear functions, are introduced in Section 2.4, and examples occur throughout the text. For instance, the coefficients of the derivative of a polynomial are a linear function of its coefficients;
>
> The function *A*+.×*ω* can be completely analyzed in terms of the array *A*. Elements of this analysis are presented in this chapter. More thorough analyses can be found in texts on linear algebra.

Just as the symbols ∘. can be used in conjunction with any primary dyadic scalar function to produce tables of values, the symbol . can be used in conjunction with any pair of primary dyadic scalar functions. For example, for matrices *A* and *B* the element at row *I*, column *J* of the matrix *A*⌈.-*B* is equal to ⌈/*A*[*I*;]-*B*[;*J*]. The function *α*⌈.-*ω* is called the ⌈.- inner product. Several functions in the text will be defined in terms of inner products other than +.×, but the +.× inner product will receive the most attention.

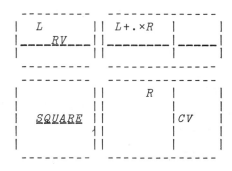

2.2 A MARKOV PROCESS

The application of the +.× inner product presented below also illustrates the idea of a Markov process or Markov chain. More information on these processes can be found in texts on probability theory.

Suppose that companies *A*,*B* and *C* are the major frozen food processors and that their combined share of the frozen food market is for all practical purposes 100 per cent. Let *PA*70, *PB*70, *PC*70 represent the proportions held by the three companies *A*, *B* and *C* during the year 1970. For example:

```
      PA70      PB70      PC70
0.60.250.15
```

The transitions in the market from one year to the next can be described by a matrix product. Suppose that during 1971 company *A* manages to keep 80 per cent of its 1970 customers and at the same time attract 10 per cent of *B*'s 1970 customers and 5 per cent of *C*'s 1970 customers. If no new customers are added to the market during 1971, then *A*'s proportion of the market during 1971 can be described by the following identity:

```
      PA71
      TA+.×P70
```

where *TA*←0.8 0.1 0.05 and *P*70←*PA*70,*PB*70,*PC*70. If during 1971 *B* keeps 85 per cent of its 1970 customers while attracting 10 per cent of *A*'s and 10 per cent of *C*'s, then *B*'s proportion of the 1971 market can be described as follows:

```
      TB←0.1  0.85  0.1

      PB71
      TB+.×P70
```

Similarly for *C*:

```
      TC←0.1  0.05  0.85

      PC71
      TC+.×P70
```

Now define the transition matrix *T*, whose rows are *TA*, *TB* and *TC*:

```
      ⎕←T←3 3ρTA,TB,TC
0.8        0.1        0.05
0.1        0.85       0.1
0.1        0.05       0.85
```

This transition matrix can be used to produce the proportions of the 1971 market held by all three companies. The transition function *TR*:*T*+.×⍵ satisfies the following identity:

```
P71←PA71,PB71,PC71

P71
TR P70
```

Thus:

```
      P70
0.6  0.25  0.15

      2⍕P71←TR P70
.51  .29  .20
```

If the transitions from 1971 to 1972 are identical to those from 1970 to 1971, then the proportions of the 1972 market are given by the same transition function *TR*:

```
      2⍕P72←TR P71
.45  .32  .24
```

In fact, as long as the transitions remain the same from year to year, this process can be continued to yield Table 2.2.1.

	PROPORTION HELD BY		
YEAR	*A*	*B*	*C*
1975	.34	.36	.29
1980	.29	.39	.32
1985	.28	.40	.32
1990	.28	.40	.32
1995	.28	.40	.32

MARKET DISTRIBUTION
TABLE 2.2.1

Evidently the percentages are stabilizing at 28, 40 and 32. This distribution of the market is called the equilibrium distribution because of the identity

```
EQ←0.28  0.40  0.32

EQ
TR EQ
```

If the distribution ever reaches equilibrium then it will remain at equilibrium as long as the year to year transitions continue unchanged. ⍠14

2.3 MATRIX PRODUCT EQUATIONS

<u>Triangular Matrices</u>. We will now consider the construction of solutions of the equation

```
∧/Y=A+.×ω
```

for a triangular matrix *A* and a <u>conformable</u> vector *Y* (that is, `(ρY)=1↑ρA`).

A matrix*A* is said to be <u>lower triangular</u> if all its elements above the diagonal are zero. Formally:

```
∧/∧/(0≠A)≤(ι1↑ρA)∘.≥ι¯1↑ρA
```

Example 1. The matrix *A* can be seen to be lower triangular by comparing the matrices to the right of it:

```
       A                 0≠A              (ι1↑ρA)ι∘.≥ι¯1↑ρA
¯6 0  0  0            1 0 0 0           1 0 0 0
¯5 9  0  0            1 1 0 0           1 1 0 0
 6 0 ¯5  0            1 0 1 0           1 1 1 0
 4 7  5 ¯3            1 1 1 1           1 1 1 1
```

Analogously, *A* is said to be an <u>upper triangular matrix</u> if

```
∧/∧/(0≠A)≤(ι1↑ρA)∘.≤ι¯1↑ρA
```

Example 2:

```
      B                  0≠B              (ι1↑ρB)∘.≤ι¯1↑ρB
5 ¯2  3  5            1 1 1 1           1 1 1 1
0 ¯1 ¯3  8            0 1 1 1           0 1 1 1
0  0  0 ¯5            0 0 0 1           0 0 1 1
0  0  0  3            0 0 0 1           0 0 0 1
```

A matrix is <u>triangular</u> if it is either upper or lower triangular.

For a matrix *A* the vector defined by the expression `0 0⍉A` is the <u>diagonal of</u> *A*. For example:

```
      ⎕←A←3 3ρι9
0 1 2
3 4 5
6 7 8

      0 0⍉A
0 4 8
```

A triangular matrix *A* is said to be <u>complete</u> if *A* is <u>square</u> and its diagonal elements are all non-zero. That is:

```
      (=/ρA)∧∧/0≠0 0⍉A
```

For example, the matrix *A* in example 1 is complete, but the matrix *B* in example 2 is not complete.

If *A* is a complete lower triangular matrix and *Y* is a conformable vector, then the equation

```
      ∧/Y=A+.×ω
```

has one and only one solution, and this solution can be produced by the <u>forward substitution algorithm</u>. For example:

```
       A                     Y
2  0  0  0           4 ¯9 33 6
3 ¯3  0  0
1  5 ¯2  0
4 ¯3 ¯2  1
```

We will represent the solution *X* of the equation `∧/Y=A+.×ω` by `X0,X1,X2,X3`. That is, the scalar `X0` represents `X[0]`, `X1` represents `X[1]`, and so on. Then in order to solve this equation we must solve the four equations:

```
 4 = (2×X0) + ( 0×X1) + ( 0×X2) + 0×X3
¯9 = (3×X0) + (¯3×X1) + ( 0×X2) + 0×X3
33 = (1×X0) + ( 5×X1) + (¯2×X2) + 0×X3
 6 = (4×X0) + (¯3×X1) + (¯2×X2) + 1×X3
```

The terms with coefficient 0 can be omitted:

```
 4 =  2×X0
¯9 = (3×X0) +  ¯3×X1
33 = (1×X0) + ( 5×X1) +  ¯2×X2
 6 = (4×X0) + (¯3×X1) + (¯2×X2) + 1×X3
```

The first equation can be solved for `X0`.

```
      ⎕←X0←4÷2
2
```

The second equation can then be solved for `X1`.

```
      ⎕←X1←(¯9-3×X0)÷¯3
5
```

The third equation can then be solved for `X2`, and so on:

```
      ⎕←X2←(33-1 5+.×ω0,X1)÷¯2
¯3

      ⎕←X3←(6-4 ¯3 ¯2+.×X0,X1,X2)÷1
7

      ⎕←X←X0,X1,X2,X3
2 5 ¯3 7
```

```
      A+.×X              Y
4 ¯9 33 6          4 ¯9 33 6
```
⌹20

The forward substitution algorithm can be described formally in terms of the matrix *A* and the vector *Y*. Continuing the preceding example:

```
      X←ι0
      X←X,(Y[0]-A[0;ι0]+.×X)÷A[0;0]
      X←X,(Y[1]-A[1;ι1]+.×X)÷A[1;1]
      X←X,(Y[2]-A[2;ι2]+.×X)÷A[2;2]
      X←X,(Y[3]-A[3;ι3]+.×X)÷A[3;3]

      X
2 5 ¯3 7
```

If *B←Y,-A* (that is, the matrix *A* bordered on the left by the vector *Y* as its first column) then the above expressions are equivalent to:

```
      X←ι0
      X←X,(B[0;ι1]+.×1,X)÷B[0;1]
      X←X,(B[1;ι2]+.×1,X)÷B[1;2]
      X←X,(B[2;ι3]+.×1,X)÷B[2;3]
      X←X,(B[3;ι4]+.×1,X)÷B[3;4]

      X
2 5 ¯3 7
```

This formal description is the basis of the definition of the function *SF* (for solve forward), which can be defined in terms of two auxiliary functions *SFA* and *SFB* as follows:

```
      SF:SFA α,-ω
      SFA:α[(ρα)[0]-1;] SFB SFA ¯1 ¯1↓α : 0=(ρα)[0] : ι0
      SFB:ω,((¯1↓(2+ρω)↑α)+.×1,ω)÷¯1↑(2+ρω)↑α

      X←ι0
      X←B[0;] SFB X
      X←B[1;] SFB X
      X←B[2;] SFB X
      X←B[3;] SFB X
```

Execution of *SFA* for a few steps will show that the function *SF* is in fact the formal solution of the equation ∧/*Y*=*A*+.×ω. For example, for the matrix *A* and the vector *Y* in the preceding example:

```
      Y SF A
2 5 ¯3 7
```

For a complete upper triangular matrix *A* and a conformable vector *Y*, the solution of the equation ∧/*Y*=*A*+.×ω can be produced by the backward substitution algorithm. For example:

```
        A                    Y
6  4  6 ¯5           ¯4 23 10 10
0  3 ¯3  4
0  0  8  1
0  0  0  5

        X←⍳0
        X←X,(Y[3]-A[3;3-⍳0]+.×X)÷A[3;3]
        X←X,(Y[2]-A[2;3-⍳1]+.×X)÷A[2;2]
        X←X,(Y[1]-A[1;3-⍳2]+.×X)÷A[1;1]
        X←X,(Y[0]-A[0;3-⍳3]+.×X)÷A[0;0]

        ⎕←X←⌽X               A+.×X
¯4 6 1 2             ¯4 23 10 10
```

The definition of the solve backward function *SB* is similar to that of *SF* and is left to the exercises. ⌹24

<u>The Primary Dyadic Function</u> *Y*⌹*A*. For a complete triangular matrix *A* and a conformable vector *Y*, the solution of the equation ∧/*Y*=*A*+.×⍵ can be produced by the appropriate substitution algorithm. This solution can also be produced by the primary dyadic function denoted *Y*⌹*A* and called <u>domino</u> or <u>quad-divide</u>. For example:

```
        A                   Y
 8  0  0  0          56 ¯14 29 10
¯5  3  0  0
 0  5  1  0
¯2 ¯1 ¯4  1

        ⎕←X←Y⌹A             A+.×X
7 7 ¯6 7             56 ¯14 29 10
```

Also:

```
        Y SF A
7 7 ¯6 7
```

The function domino will also produce the solution of the equation ∧/*Y*=*A*+.×⍵ for a nontriangular matrix *A*. For example:

```
        A                   Y
1  1  1              4 30 72
1  3  9
1  5 25

        ⎕←X←Y⌹A             A+.×X
¯3 5 2               4 30 72
```

Not all square matrices are valid right arguments of domino. By definition, a square matrix *A* is a valid right argument of domino if and only if for every vector *Y* the equation ∧/*Y*=*A*+.×⍵ has one and only one solution. For example, a square triangular matrix is a member of the right argument domain of domino if and only if it is complete (see Exs 2.22 and 2.23). ⊞27

2.4 EXTENDED DOMAINS OF MONADIC FUNCTIONS

One of the important characteristics of the primary scalar functions is that they apply element-by-element to argument arrays. Many of the primary non-scalar functions enjoy a similar characteristic.

For example, consider the function ⌽⍵. When applied to a vector *V*, this function reverses the order of the elements of *V*:

```
      ⌽1 2 3 4 5
5 4 3 2 1
```

When applied to a matrix *M*, this function reverses the order of the columns of *M*. For example:

```
      ⎕←M←4 4⍴⍳16                 ⌽M
 0  1  2  3                3  2  1  0
 4  5  6  7                7  6  5  4
 8  9 10 11               11 10  9  8
12 13 14 15               15 14 13 12
```

In effect, when applied to a matrix *M*, ⌽⍵ reverses the elements of each row of *M*. Continuing the preceding example:

```
      I←1
      ⌽M[I;]                     (⌽M)[I;]
7 6 5 4                    7 6 5 4
```

Note that row *I* of *M* is *M*[*I*;] or *M*[*I*;0 1 2 3]. In words, to select row *I* of *M* we choose the index along the first axis to be *I* and take all indices along the last axis. Thus we say that the rows of *M* are the vectors along the last axis of *M* and also that the function ⌽⍵ <u>applies along the last axis</u> of its arguments. Formally, a monadic function *F* applies along the last axis of matrix arguments *M* if for each index *I*:

```
      F M[I;]                                  [2.4.1]
      (F M)[I;]
```

Thus *F* applies along the last axis of matrix arguments if *F* <u>commutes</u> with selection along the first axis.

As a second example consider the function ⊖ω. As with the function ⌽ω, this function reverses the order of the elements of an argument vector *V*. For example:

```
      ⊖1 2 3 4 5
5 4 3 2 1
```

However, when applied to a matrix *M*, this function reverses the order of the rows of *M*. For example:

```
      M                                         ⊖M
 0  1  2  3                              12 13 14 15
 4  5  6  7                               8  9 10 11
 8  9 10 11                               4  5  6  7
12 13 14 15                               0  1  2  3
```

In effect, this function reverses the elements of each column of an argument matrix *M*. Continuing the preceding example:

```
      I←0
      ⊖M[;I]                                    (⊖M)[;I]
12 8 4 0                                 12 8 4 0
```

Just as we say that the function ⌽ω applies along the last axis of its arguments, we say that ⊖ω applies along the first axis.

The function ⊖ω is equivalent to ⌽[0]ω, and in the latter notation the integer inside the square brackets indicates the axis along which the function is to be applied.

It is often useful to extend the domains of secondary functions in ways similar to its extensions of primary functions illustrated above. For example, the function *EDIFF* defined below applies *DIFF* along the first axis of argument matrices.

```
      EDIFF:(DIFF ω[;0]),EDIFF 0 1↓ω : 0=¯1↑ρω : 1 0↓ω
```

```
      A                                         ⎕←Y←EDIFF A
  0   1   4   9                          16 24 32  40
 16  25  36  49                          48 56 64  72
 64  81 100 121                          80 88 96 104
144 169 196 225
```

```
      I←2
      Y[;I]                                     DIFF A[;I]
32 64 96                                 32 64 96
```

A function that applies *DIFF* along the last axis of argument matrices can be defined in a similar way. It can also be defined directly in terms of *EDIFF* and the primary monadic function denoted by ⍉ and called transpose. The transpose function has the effect of interchanging the rows and columns of argument matrices. Formally:

```
(⍉M)[I;J]
M[J;I]
```

For example:

```
      3 3⍴⍳9                           ⍉3 3⍴⍳9
0 1 2                            0 3 6
3 4 5                            1 4 7
6 7 8                            2 5 8
```

Evidently, the function ⍉*EDIFF* ⍉⍵ applies *DIFF* along the last axis. ⌹32

In general, a monadic function *F* which applies to vectors is applied along the first axis of matrices by its extended domain function

```
EF:(F ⍵[;0]),EF 0 1↓⍵ : 0=¯1↑⍴⍵ : SF ⍵
```

where *SF* is a <u>shape function</u> of *F*. That is, if ⍴*F A* is *N* when ⍴*A* is *M*, then ⍴*SF B* is *N*,0 when ⍴*B* is *M*,0. For example, the shape functions of *DIFF* and *SUM*:0,+\⍵ are *SDIFF*:1 0↓⍵ and *SSUM*:(-1 0+⍴⍵)↑⍵.

<u>Linear Functions</u>. The monadic function *F* is said to <u>distribute over</u> the dyadic function *D* if

```
F (X D Y)
(F X) D (F Y)
```

For example, the function *F*:*A*×⍵ distributes over +:

```
      A←3
      F 5+6                            (F 5)+F 6
33                               33
```

The function *G*:*L*∧⍵ distributes over ∨:

```
      L←1
      G 1∨0                            (G 1)∨G 0
1                                1
```

For a positive scalar *A* the function *H*:*A*×⍺ distributes over ⌈:

```
      A←3
      H 4⌈5                            (H 4)⌈H 5
15                               15
```

The monadic function *F* is a <u>linear function</u> if it distributes over +. For example, the function in the first example above is a linear function. Also, the function called *DIFF* is a linear function (Ex 1.17). Finally, *A*+.×ω and ω+.×*B* are linear functions (Exs 2.9 through 2.12). ⍰33

For a non-negative integer *N* the function value *ID N*, where *ID*:(ιω)∘.=ιω, is called an identity matrix because of the following identities for vectors *X*:

```
        X                                          X
        (ID ρX)+.×X                                X+.×ID ρX
```

For example:

```
        X
5 8 ¯4 10

        ID ρX
1 0 0 0
0 1 0 0
0 0 1 0
0 0 0 1

        (ID ρX)+.×X                                X+.×ID ρX
5 8 ¯4 10                                5 8 ¯4 10
```

A linear function *F* whose domain consists of vectors can be represented as a +.× inner product by applying *F* along either axis of an identity matrix. Specifically, for vectors *X*:

```
        F X                                                    [2.4.2]
        (EF ID ρX)+.×X
```

For example, since *DIFF* is a linear function it can be represented as a +.× inner product:

```
        ⎕←M←EDIFF ID 4
¯1  1  0  0
 0 ¯1  1  0
 0  0 ¯1  1

        X←13 11 18 16

        DIFF X                                     M+.×X
¯2 7 ¯2                                  ¯2 7 ¯2
```

Also, for vectors *Y*:

```
        F Y                                                    [2.4.3]
        Y+.×⍉EF ⍉ID ρY
```
⍰36

It is usual to define a function *F* to be linear if *F* distributes over + and if for scalars *A*, *F* commutes with *A*×ω. Formally:

```
F A×ω
A×F ω
```

However, for linear functions of practical interest the latter identity follows from distributivity over +, and so for our purposes it is not necessary to include it in the definition of linearity.

The Determinant. To each square matrix *A* there is associated a constant called the determinant of *A* which is not zero if and only if *A* is a valid right argument of domino. We will now define a determinant function called *DET*. It will be defined recursively in terms of determinants of minors; a minor of a matrix *A* is obtained by eliminating one row and one column of *A*. This definition of *DET* has been chosen because you may be familiar with the definition of the determinant of a matrix of size 3 3 in terms of determinants of minors. Otherwise, the definition is not of fundamental importance in this text, and *DET* can be used as if it were a primary function, such as domino.

If ∧/1=ρ*A* then *DET A* is the element of *A*, that is, *A*[0;0]. If ∧/1<ρ*A* then *DET A* is defined in terms of the functions *MINORS* and *IMINORS* (for indices to produce minors):

```
IMINORS:(-ιω)⌽0 1↓(ιω)⌽(ω,ω)ριω
MINORS:0 0 1↓ω[IMINORS 1↑ρω;]
```

The value *M*←*MINORS A* is a three dimensional array whose *I*th plane *M*[*I*;;] is the minor obtained by eliminating row *I* and column 0 of *A*. For example:

```
      ⎕←A←4 4ρι16
 0  1  2  3
 4  5  6  7
 8  9 10 11
12 13 14 15

      M←MINORS A
      M[2;;]                            A[0 1 3;1 2 3]
 1  2  3                           1  2  3
 5  6  7                           5  6  7
13 14 15                          13 14 15
```

If we apply *DET* to matrices along the last two axes of *M*, the result is a vector of determinants of minors of *A*. The function *DET* is defined as follows:

```
DET:ω[;0]-.×E̲DET MINORS ω : ∧/1=ρω : ω[0;0]
```

```
      EDET:(DET ω[0;;]),EDET 1 0 0↓ω : 0=1↑ρω : ι0
```

For example, the matrix

```
      ⎕←A←3 3ρ1 1 1 1 2 4 1 3 9
1 1 1
1 2 4
1 3 9
```

is a valid right argument of domino and *DET A* is not zero.

```
      DET A
2
```

Also, the matrix 3 3ρι9 is not a valid right argument of domino and *DET* 3 3ρι9 is zero.

```
      DET 3 3ρι9
0
```

Applications of the determinant will occur throughout the text. ⌹46

2.5 THE GEOMETRY OF VECTORS

We conclude this chapter with a brief review of analytic geometry.

Vectors *V* for which 2=ρ*V* represent points in a coordinate plane. They also represent directed line segments in a coordinate plane, as illustrated in Figure 2.5.1, where the tail of a directed line segment is located at the origin and the head is located at the coordinate point represented by a vector.

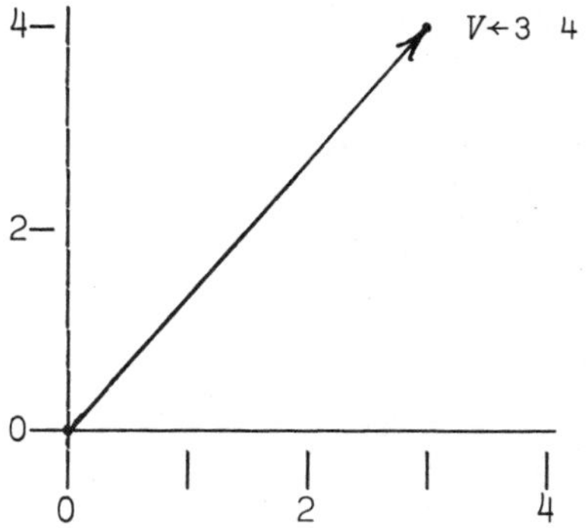

VECTORS AS DIRECTED LINE SEGMENTS
FIGURE 2.5.1

For the function *NORM*:(+/ω⋆2)⋆0.5 the scalar *NORM V* is called the norm of the vector *V* and *NORM V* is the distance from the point represented by to the origin, as well as the length of the directed line segment represented by *V*. For the function *DIRECTION*:ω÷*NORM* ω the vector *DIRECTION V* is called the direction of the vector *V*; note that the norm of the direction of *V* is 1. The directed line segment represented by *DIRECTION V* lies on the one represented by *V*, as illustrated in Figure 2.5.2. For the functions *NORM* and *DIRECTION* we have the identity

```
V                                              [2.5.3]
(NORM V)×DIRECTION V
```

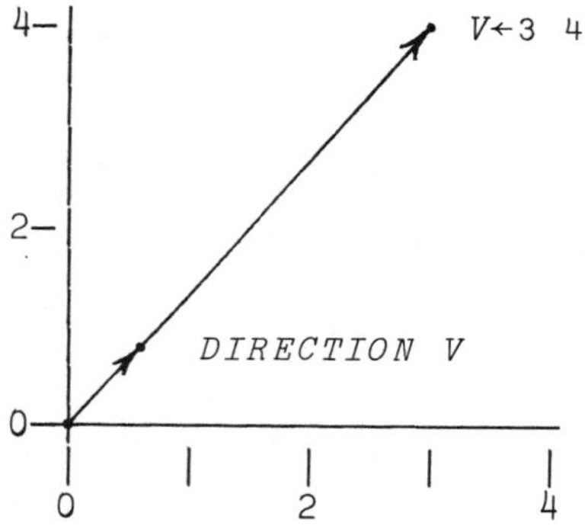

DIRECTION VECTORS
FIGURE 2.5.2

Also, for a scalar *S* and a vector *V*:

```
NORM S×V           DIRECTION S×V                [2.5.4]
(|S)×NORM V        (×S)×DIRECTION V
```

In particular, if *S*>0 then the directed line segment represented by *S*×*V* lies on the one represented by *V*, while if *S*<0 then the directed line segment represented by *S*×*V* points in the opposite direction of the one represented by *V*, as in Figure 2.5.5.

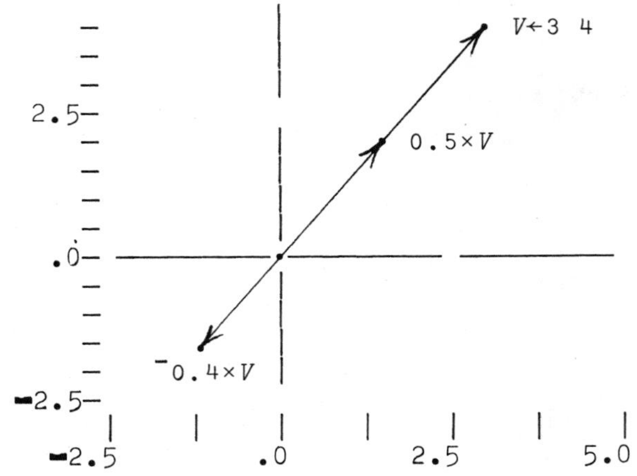

SCALAR MULTIPLES OF A VECTOR
FIGURE 2.5.5

Finally, the vector 0.5×*V*+*W* represents the midpoint of the line segment joining the points represented by *V* and *W*, as in Figure 2.5.6. ⌸49

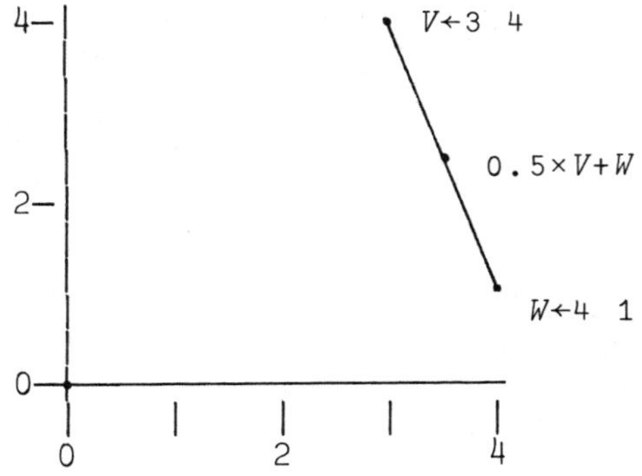

MIDPOINT OF A LINE SEGMENT
FIGURE 2.5.6

3
Polynomials

3.1 INTRODUCTION

A polynomial is a weighted sum of the functions ω⋆*N* for non-negative integers *N*. For example, the function *P*:3+(2×ω)+4×ω⋆2 is a polynomial. For a vector *C* the expression +/*C*×ω⋆⍳⍴*C* also defines a polynomial; the vector *C* is called the coefficient vector of the polynomial. For example, the function *P* above can also be defined as *P*:+/3 2 4×ω⋆0 1 2.

Even though +/*C*×ω⋆⍳⍴*C* is a scalar function, unlike the primary scalar functions it does not extend element-by-element to array arguments. However, this deficiency is easily overcome. The following are equivalent definitions of the function +/*C*×ω⋆⍳⍴*C* for scalar arguments:

```
      C+.×ω⋆⍳⍴C
      (ω⋆⍳⍴C)+.×C
```

The function ω∘.⋆⍳⍴*C* is equivalent to ω⋆⍳⍴*C* for scalars ω and also applies element-by-element to arrays ω, producing a matrix whose rows are the powers of the elements of ω; that is, the *I*th row is ω[*I*]⋆⍳⍴*C*. For example:

```
      2 5∘.⋆⍳3                2⋆⍳3                5⋆⍳3
1 2 4                   1 2 4               1 5 25
1 5 25
```

If the expression ω⋆⍳⍴*C* in (ω⋆⍳⍴*C*)+.×*C* is replaced by ω∘.⋆⍳⍴*C*, the resulting function will apply element-by-element to an array ω. For example, compare the results of the functions *P*:+/3 2 4×ω⋆0 1 2 and *Q*:(ω∘.⋆0 1 2)+.×3 2 4:

```
      Q 5 ¯3 0              P 5              P ¯3              P 0
113  33  3              113               33                3
```

Values of the function ω∘.⋆⍳α are called Vandermonde arrays (for the 18th century Dutch mathematician A.T. Vandermonde), and consequently we define *VAN*:ω∘.⋆⍳α.

Just as a vector can be thought of as the coefficient vector of a polynomial, a matrix can be thought of as a family of coefficient vectors, where each column is a coefficient vector. For example, the matrix

```
      ⎕←M←3 3⍴2 4 7 ¯3 0 ¯1 1 2 1
 2  4  7
¯3  0 ¯1
 1  2  1
```

represents the family of coefficient vectors 2 ¯3 1 and 4 0 2 and 7 ¯1 1. The function ((1↑ρ*M*) *VAN* ω)+.×*M* will produce values of every polynomial in a family *M*. For example, for the above matrix *M*:

```
      ((1↑ρM) VAN 5)+.×M
12 54 27
      +/2 ¯3 1×5*0 1 2
12
      +/4 0 2×5*0 1 2
54
      +/7 ¯1 1×5*0 1 2
27
```

Using the above discussion as a guide, we define *POLY*:((1↑ρα) *VAN* ω)+.×α. If *C* is a vector then *C POLY* ω is the polynomial with coefficient vector *C*. If *M* is a matrix then the function *M POLY* ω produces values of all polynomials whose coefficient vectors are the columns of *M*. Continuing the above example:

```
      M POLY 5                  M POLY 5 6
12 54 27                  12 54 27
                          20 76 37
```

It is clear from its definition that the function *POLY* is a linear function of its left argument; we often say, somewhat loosely, that polynomials are linear functions of their coefficients. This fact will be exploited throughout the text.

The <u>degree</u> of the polynomial with coefficient vector *C* is defined to be the largest integer *I* for which 0≠*C*[*I*]. That is, *DEG*:⌈/(ιρα)×0≠α. For example:

```
      C←1 0 2 4 3 0 0 0
      DEG C
4
```

More generally, if *DEGREE*:(ι1↑ρα)⌈.×0≠α and *M* is a matrix, then *DEGREE M* is a vector of degrees of the polynomials whose coefficient vectors are the columns of *M*.

The <u>order</u> of the polynomial with coefficient vector *C* is defined to be ¯1+ρ*C*, and we define *ORDER*:¯1+1↑ρω. The coefficient vector *C* is said to be of <u>normal form</u> if *ORDER C* is equal to *DEGREE C*. ⌹1

<u>Functions on Polynomials</u>. Many functions applied to polynomials can be described as functions applied to coefficient vectors, and this is true of most of the functions introduced in this text. In particular, the derivative of the polynomial with coefficient vector *C* is the polynomial with coefficient vector 1↓*C*×ιρ*C*. For example, the polynomial *P*:1 ¯4 5 *POLY* ω is equivalent to

1+(¯4×ω)+5×ω*2, and the derivative of the latter expression is ¯4+10×ω. Therefore the derivative of *P* is *DP*:¯4 10 *POLY* ω, and ¯4 10 is 1↓1 ¯4 5×ι3. In general, the derivative of the polynomial with coefficient vector *C* is the polynomial with coefficient vector *DERIV C*, where *DERIV*:1↓α×ιρα. ⌹6

Analogously, for the function *INTEG*:α,ω÷1+ιρω and a scalar *S*, the polynomial with coefficient vector *S INTEG C* is an integral of the polynomial with coefficient vector *C*. To see that this is true we must show that *DERIV S INTEG C* is *C*:

```
DERIV S INTEG C
1↓(S,C÷1+ιρC)×ιρS,C÷1+ιρC
1↓(S,C÷1+ιρC)×ιρS,C
1↓(S,C÷1+ιρC)×0,1+ιρC
1↓0,C
C                                   ⌹17
```

Just as the derivative and integrals of polynomials are also polynomials, so are the sum, product, and composite of two polynomials. Thus each of (*A POLY* ω)+*B POLY* ω and (*A POLY* ω)×*B POLY* ω and *A POLY B POLY* ω can be expressed as *C POLY* ω for an appropriate coefficient vector *C*. Moreover, *C* is evidently a function of *A* and *B* in each case. That is, there are functions *PLUS*, *TIMES*, and *COMP* for which the following identities are valid:

```
(A POLY ω)×B POLY ω(A POLY ω)×B POLY ω
(A PLUS B) POLY ω(A TIMES B) POLY ω

A POLY B POLY ω
(A COMP B) POLY ω
```

Unless *A POLY* ω is of degree 0, the reciprocal function ÷*A POLY* ω is not a polynomial. However, its values can be approximated by values of polynomials. Coefficient vectors of the approximating polynomials are produced by the function to be called *REC*. Specifically, if 0≠*A*[0] then *N REC A* is the coefficient vector of a polynomial of degree *N* whose values are approximately equal to those of ÷*A POLY* ω, at least for arguments near 0.

The algebra of polynomials consists of the functions *PLUS*, *TIMES*, *REC*, and *COMP*, and detailed discussions of their definitions are presented in the next four sections. These functions will be used to define an algebra of approximating polynomials for non-polynomial functions in Chapter 4.

The importance of polynomials is based on the following two facts:

> Polynomials can be analyzed in detail without introducing non-polynomial functions, such as fractional powers;
>
> Most non-polynomial functions of practical interest can be approximated as closely as desired by polynomials.

Both the analysis of polynomials and their use in approximating and constructing non-polynomial functions will be studied in this text.

3.2 THE SUM OF POLYNOMIALS

The function *(A POLY ω)+B POLY ω* is a polynomial whose coefficient vector can be produced as a function of *A* and ⊥. That is, there is a function ⋆⎕↓⌈ for which *(A POLY ω)+B POLY ω* is equivalent to *(A PLUS B) POLY ω*. *PLUS* can be defined as

```
      PLUS:(((ρα)⌈ρω)↑α)+((ρα)⌈ρω)↑ω
```

For example:

```
      A←4 2 1
      B←5 0 2 ¯3 1
      C←A PLUS B
      C
9 2 3 ¯3 1

      (A POLY 5)+B POLY 5             C POLY 5
344                             344
```

The definition of *PLUS* will be discussed in the exercises. ⌸22

3.3 THE PRODUCT OF POLYNOMIALS

The product of two polynomials is also a polynomial and its coefficient vector can be produced as a function of *A* and *B*. Precisely, for

```
      TIMES:+⌿(-ιρα)⌽α∘.×ω,0×1↓α
```

the function *(A POLY ω)×B POLY ω* is equivalent to the function *(A TIMES B) POLY ω*. For example:

```
      A←2 ¯1 4
      B←1 0 ¯2 1
      C←A TIMES B
      C
2 ¯1 0 4 ¯9 4

      (A POLY 2)×B POLY 2             C POLY 2
16                              16
```

To understand the definition of *TIMES*, consider the following sequence of identities for the product of the polynomials with coefficient vectors *A* and *B*:

```
      (+/A×ω⋆⍳ρA)×+/B×ω⋆⍳ρB
      +/+/(A×ω⋆⍳ρA)∘.×B×ω⋆⍳ρB
      +/+/(A∘.×B)×(ω⋆⍳ρA)∘.×ω⋆⍳ρB
      +/+/(A∘.×B)×ω⋆(⍳ρA)∘.+⍳ρB
```

The elements of the table `(⍳ρA)∘.+⍳ρB` are integers between `0` and `¯2+(ρA)+ρB` that are identical along the upper-right to lower-left diagonals. For example:

```
      (⍳3)∘.+⍳4
0 1 2 3
1 2 3 4
2 3 4 5
```

Consequently, if the elements of the table `A∘.×B` are summed along the upper-right to lower-left diagonals and the resulting vector is denoted by *P*, the above sequence of identities can be continued as follows:

```
      +/P×ω⋆⍳¯1×(ρA)+ρB
      +/P×ω⋆⍳ρP
```

Evidently, *P* is a coefficient vector of the product of the polynomials with coefficient vectors *A* and *B*. The function *TIMES* is defined to produce sums along the upper-right to lower-left diagonals of tables `A∘.×B`. For example:

```
      A←3 1 4
      B←2 5 ¯1 1

      A∘.×B
6  15  ¯3   3
2   5  ¯1   1
8  20  ¯4   4

      A∘.×B,1↓0×A
6  15  ¯3   3   0   0
2   5  ¯1   1   0   0
8  20  ¯4   4   0   0

      (-⍳ρA)⌽A∘.×B,1↓0×A
6  15  ¯3   3   0   0
0   2   5  ¯1   1   0
0   0   8  20  ¯4   4

      +⌿(-ρA)⌽A∘.×B,1↓0×A
6  17  10  22  ¯3 4

      DEGREE A TIMES B                    (DEGREE A)+DEGREE B
5                                  5
```

The relation suggested by the last results can be expressed as follows:

```
      DEGREE α TIMES ω                                    [3.3.1]
      (DEGREE α)+DEGREE ω
```
⌹28

It has been pointed out that a matrix *C* can be thought of as a family of polynomial coefficient vectors, the coefficient vectors being the columns of *C*. For example:

```
      C←3 3ρ2 ¯1 0 4 5 ¯7 2 0 1
      C
 2 ¯1  0
 4  5 ¯7
 2  0  1
```

is the family of coefficient vectors `2 4 2` and `¯1 5` and `0 ¯7 1`. The right argument domain of *TIMES* is extended to apply to families of coefficient vectors by the following functions:

```
   ETIMES:(α TIMES ω[;0]),α ETIMES 0 1↓ω : 0=¯1↑ρω : α STM ω
   STM:((ρω)+(¯1+ρα),0)ρω
```

For example:

```
      A←2 0 5 2 ETIMES C
      A[;1]
¯2 10 ¯5 23 10 0
      2 0 5 2 TIMES C[;1]
¯2 10 ¯5 23 10 0
```

Reduction by *TIMES* can be applied to families of coefficient vectors in the same way that reduction by times is applied to vectors. According to the definition of `×/ω`, for a vector *V* of length at least `1`:

```
      ×/V
      V[0]××/1↓V
```

Also, `×/ι0` is `1`. Thus reduction by times can be recursively defined as follows:

```
      TR:ω[0]×TR 1↓ω : 0=ρω : 1
```

Similarly, the function

```
      PTR:α[;0] TIMES PTR 0 1↓α : 0=¯1↑ρα : 1ρ1
```

produces the reduction by *TIMES* of a family of coefficient vectors. Continuing the preceding example:

```
      PTR C
0 14 ¯44 ¯120 ¯52 10 0
```

```
      C[;0] TIMES C[;1] TIMES C[;2]
0 14 ¯44 ¯120 ¯52 10 0
```
⍠38

Analogously, the augmented *TIMES* scan of a family of coefficient vectors is produced by the function

```
      ATS:(ATS 0 ¯1↓ω) APPEND PTR ω : 0=¯1↑ρω : 1 1ρ1
```

where

```
      APPEND:(((ρω),¯1↑ρα)↑α),ω
```

Continuing the above example:

```
      PTS C
1     2    ¯2     0
0     4     6    14
0     2    18   ¯44
0     0    10  ¯120
0     0     0   ¯52
0     0     0    10
0     0     0     0
```

3.4 THE COMPOSITION OF POLYNOMIALS

For polynomials *F:A POLY ω* and *G:B POLY ω* the composite function *H:F G ω* is also a polynomial. For example, let *A*←1 2 3. Then for *H ω* we have the sequence of identities:

```
      H ω
      F G ω
      1+(2×G ω)+3×(G ω)*2
      1+(2×B POLY ω)+3×(B TIMES B) POLY ω
```

Thus if *M* is a matrix with three columns whose columns 0,1 and 2 are coefficient vectors of polynomials equivalent to (1ρ1) *POLY ω* , *B POLY ω* , and (*B TIMES B*) *POLY ω* respectively, then *H ω* is equivalent to (*M*+.×1 2 3) *POLY ω*. Evidently, *M* can be defined as the augmented *TIMES* scan of a two-column matrix, both of whose columns are *B*.

In general, a coefficient vector of the composite of two polynomials is produced by the function

```
      COMP:(ATS ω∘.+0×1↓α)+.×α
```

For example:

```
      A←1 ¯3 2
      B←1 0 ¯3 1
      C←A COMP B
      C
0 0 ¯3 1 18 ¯12 2
```

```
      A POLY B POLY 5                         C POLY 5
5050                                    5050
```

Note that:

```
      DEGREE A COMP B                                [3.4.1]
      (DEGREE A)×DEGREE B
```

The definition of *COMP* will be examined in the exercises. ⌹44

3.5 RECIPROCAL POLYNOMIALS

Although the reciprocal of a polynomial of degree greater than zero is not a polynomial it can be approximated by polynomials. For example, consider the polynomial *P*:1-ω for which we have the familiar identity

```
      ÷P ω
      (+/ω*⍳N)+(ω*N)÷P ω
```

or equivalently, for

```
      RCP:(αρ1) POLY ω
```

and

```
      RMP:ω*α
```

we have

```
      ÷P ω                                           [3.5.1]
      (N RCP ω)+(N RMP ω)÷P ω
```

For arguments *X* such that 1>|*X* the polynomials *N RMP* ω assume values closer and closer to zero as *N* is assigned larger and larger values. For example:

```
      2 RMP 0.1
0.001
      5 RMP 0.1
0.000001
```

Evidently (*N RMP* ω)÷*P* ω also assumes values close to 0. Thus for these arguments and large values of *N*, the values of the polynomials *N RCP* ω are approximately equal to the values of the reciprocal function ÷*P* ω. Note that, according to 3.5.1, if *ONE*:1+0×ω then

```
      ONE ω
      ((N RCP ω)×P ω)+N RMP ω
```

The polynomial *N RCP* ω is called a reciprocal polynomial of the polynomial *P* and *N RMP* ω is called its remainder.

In general, for a coefficient vector *C*, if *Q* and *R* are coefficient vectors for which

```
∧/0=(ρQ)↑R                                    [3.5.2]
```

and

```
ONE ω
((C POLY ω)×Q POLY ω)+R POLY ω
```

or equivalently

```
ONE ω                                         [3.5.3]
(R PLUS C TIMES Q) POLY ω
```

then *Q POLY ω* is called a reciprocal polynomial of *C POLY ω* and *R POLY ω* is called its remainder.

In order to produce the coefficient vector *Q* of a reciprocal polynomial of the polynomial *C POLY ω*, first note that the function (*K*↑1) *POLY ω* is equivalent to *ONE ω* for a positive integer *K*. In particular, *ONE ω* is equivalent to the zeroth degree polynomial ((ρ*R PLUS C TIMES Q*)↑1) *POLY ω*, and therefore 3.5.3 is equivalent to

```
(ρR PLUS C TIMES Q)↑1                         [3.5.4]
R PLUS C TIMES Q
```

Since the first ρ*Q* elements of the remainder vector *R* are zero we have the sequence of identities:

```
(ρQ)↑1
(ρQ)↑(ρR PLUS C TIMES Q)↑1
(ρQ)↑R PLUS C TIMES Q               3.5.4
(ρQ)↑C TIMES Q                      3.5.2
```

We now have a simple identity involving *C* and *Q*, namely:

```
(ρQ)↑1
(ρQ)↑C TIMES Q
```

Since (ρα)↑*C TIMES* α is a linear function it can be represented as a +.× inner product. Specifically:

```
(ρQ)↑C TIMES Q
((2ρρQ)↑C ETIMES ID ρQ).×Q
```

Combining the last two identities yields

```
(ρQ)↑1
((2ρρQ)↑C ETIMES ID ρQ).×Q
```

The result of (2ρω)↑α *ETIMES ID* ω is a lower triangular.

For example:

```
        5 5↑3 ¯4 5 ETIMES ID 5
 3  0  0  0  0
¯4  3  0  0  0
 5 ¯4  3  0  0
 0  5 ¯4  3  0
 0  0  5 ¯4  3
```

Consequently vectors *Q* for which the preceding identity is valid can be produced by the forward substitution algorithm. That is:

```
        ∧/Q=((ρQ)↑1) SF (2ρρQ)↑C ETIMES ID ρQ
```

Thus we define the polynomial reciprocal function

```
        REC:(α↑1) SF (α,α)↑ω ETIMES ID α
```

For example, we have seen that the polynomials with coefficient vectors *N*ρ1 are reciprocal polynomials of 1-ω, or equivalently, 1 ¯1 *POLY* ω. Therefore we must have the identity

```
        Nρ1
        N REC 1 ¯1
```

For example:

```
        5 REC 1 ¯1 8 REC 1 ¯1
1 1 1 1 11 1 1 1 1 1 1
```

Note that once we can produce coefficient vectors *Q* of reciprocal polynomials of *C POLY* ω, the coefficient vectors of their remainders are simply

```
        ((ρC TIMES Q)↑1)-C TIMES Q
```

(Identity 3.5.4). Continuing the previous example:

```
        Q←7 REC 1 ¯1
        R←((ρ1 ¯1 TIMES Q)↑1)-C TIMES Q
        R
0 0 0 0 0 0 0 1
```

⌹52

You may have noticed a relationship between the definition of the function *REC* and the polynomial division algorithm developed in Exercises 3.35-36. To illustrate this relationship, let *A*←2 3 ¯1 4, suppose that *B* is a vector of length 6, and consider the problem of determining a vector *Q* for which

```
        ∧/B=A TIMES Q                                    [3.5.5]
```

According to Exercise 3.35, ⍴*Q* is 3, and consequently *A TIMES Q* is equal to *M*+.×*Q*, where *M* is *A ETIMES ID* 3.

```
      ⎕←M←A ETIMES ID 3
 2  0  0
 3  2  0
¯1  3  0
 4 ¯1  3
 0  4 ¯1
 0  0  4
```

Consequently, 3.5.5 is equivalent to

```
      ∧/B=M+.×Q                                          [3.5.6]
```

Since *M* has more rows than columns, the latter equation does not necessarily have a solution. However, in this example, if *I* is a vector of any three distinct row indices of *M*, then the equation

```
      ∧/B[I]=M[I;]+.×Q
```

does have a solution *Q*. In particular, if *I* is 3 4 5 then the solution *Q* is the quotient in the polynomial division algorithm, while if *I* is 0 1 2 and *B* is 6↑1 the solution *Q* is a coefficient vector of a reciprocal polynomial of *A POLY* ⍵. In either case, Equation 3.5.5 does not have a solution, and the remainder is the measure of the difference between *B* and *A TIMES Q*.

3.6 OTHER EXPRESSIONS FOR POLYNOMIALS

We have defined polynomials to be weighted sums of the functions ⍵⋆*N* for non-negative integers *N*. This definition can be too restrictive in applications, where it is often useful to express polynomials as weighted sums of the functions (⍵-*A*)⋆*N* for non-negative integers *N* and a constant *A*. These weighted sums will be called polynomials centered at *A*, or simply polynomials at *A*. Thus the functin *C POLY* ⍵-*A* is called the polynomial with coefficient vector *C* centered at *A*. In particular, *C POLY* ⍵ is the polynomial with coefficient vector centered at 0.

If the polynomials *C POLY* ⍵-*A* and *D POLY* ⍵ are equivalent functions, then *D* can be produced in terms of *C* and *A*, and *C* in terms of *D* and *A*. For that purpose consider the function ⍵-*A*, which is a polynomial centered at *A*, with coefficient vector 0 1. It is also a polynomial centered at 0, with coefficient vector (-*A*),1. Therefore

```
      C POLY ⍵-A
      C POLY ((-A),1) POLY ⍵
      (C COMP (-A),1) POLY ⍵
```

Similarly, the function *ω* is a polynomial centered at 0, with coefficient vector 0 1. It is also a polynomial centered at *A*, with coefficient vector *A*,1. Therefore

```
D POLY ω
D POLY (A,1) POLY ω-A
(D COMP A,1) POLY ω-A
```

Thus if *C* is the coefficient vector of a polynomial centered at *A*, then *C COMP* (-*A*),1 is the coefficient vector of an equivalent polynomial, centered at 0. Also, if *D* is the coefficient vector of a polynomial centered at 0, then *D COMP A*,1 is the coefficient vector of an equivalent polynomial, centered at *A*.

Derivatives and integrals of polynomials centered at *A* are produced as before. That is, the derivative of *C POLY ω-A* is (*DERIV C*) *POLY ω-A*, and for each scalar *S* the polynomial (*S INTEG C*) *POLY ω-A* is an integral of *C POLY ω-A*.

The algebra of polynomials extends to polynomials centered at *A*. Evidently:

```
(B POLY ω-A)+C POLY ω-A
(B PLUS C) POLY ω-A
```

and

```
(B POLY ω-A)×C POLY ω-A
(B TIMES C) POLY ω-A
```

The extension of polynomial composition is not quite so straightforward, but is based on a simple fact. Namely, if *G*:*C POLY ω-A*, then *G A* is *C*[0]. Consequently:

```
(G ω)-G A
(0,1↓C) POLY ω-A
```

and if *F*:*B POLY ω-G A*, then

```
F G ω
B POLY (G ω)-G A
B POLY (0,1↓C) POLY ω-A
(B COMP 0,1↓C) POLY ω-A
```

In words, if *G* is the polynomial with coefficient vector *C* centered at *A*, and if *F* is the polynomial with coefficient vector *B* centered at *G A*, then the composite function *F G ω* is the polynomial with coefficient vector *B COMP* 0,1↓*C*, centered at *A*.

Reciprocal polynomials illustrate the importance of expressing polynomials as weighted sums of the functions (*ω-A*)⋆*N*. For example, the reciprocal polynomials (*N REC* 1 ¯*A*) *POLY ω* provide approximate values of the reciprocal function ÷1-*ω* only for arguments *X* for which

1>|*X*. To produce approximate values of ÷1-⍵ for arguments near *A*, where 1≤|*A*, first express 1-⍵ as a polynomial centered at *A*. Namely:

```
1-⍵
(1 ¯1 COMP A,1) POLY ⍵-A
```

Then the functions (*N REC* 1 ¯1 *COMP A*,1) *POLY* ⍵-*A* are called reciprocal polynomials of 1-⍵ centered at *A*, and these polynomials provide approximate values of ÷1-⍵ for arguments near *A*.

In general, the polynomials (*N REC C COMP A*,1) *POLY* ⍵-*A* are called the reciprocal polynomials of *C POLY* ⍵ centered at *A*, and the values of these polynomials approximate the values of the reciprocal function ÷*C POLY* ⍵ for arguments near *A*.

3.7 FITTING POLYNOMIALS

Up to now we have been mainly concerned with the algebra of polynomials. In this section one of the important applications of polynomials is introduced, which is often referred to as curve fitting. The idea is that for any *N* distinct points in a coordinate plane which lie on the graph of a function *G* there is a polynomial of degree at most *N*-1 whose graph passes through all *N* points. Thus the values of *G* and the polynomial agree for *N* arguments and, while there is no guarantee that their values agree elsewhere, it is reasonable to expect that they are approximately equal for other arguments. Such a polynomial is called a fitting polynomial for the function *G*. The problem we are concerned with here is defining a function that produces coefficient vectors of fitting polynomials. Applications will occur throughout the text.

Every polynomial of degree *N*-1 has a coefficient vector of length *N*; we are interested in producing the coefficient vector *C* of length *N* for which

```
∧/(G X)=C POLY X                                  [3.7.1]
```

for a specific vector of arguments *X* of length *N*. Then *C POLY* ⍵ is a fitting polynomial for *G* ⍵. Note that the values of *C POLY* ⍵ and *G* ⍵ will not necessarily agree for arguments other than the elements of *X*. However, the values of *C POLY* ⍵ and *Y* ⍵ may be approximately the same, at least in the interval from ⌊/*X* to ⌈/*X*.

According to the definition of the function *POLY*, 3.7.1 is equivalent to

```
∧/(G X)=((⍴C) VAN X)+.×C
```

and, since (⍴C)=⍴X,

```
      ∧/(G X)=((⍴X) VAN X)+.×C                    [3.7.2]
```

Thus to produce the coefficient vector C we must solve 3.7.2. The solution of this equation can be produced by the function α⌹ω if the matrix (⍴X) VAN X is a valid right argument.

The vector V is said to be lean if no two elements of V are equal. Formally, V is lean if ∧/1=+/V∘.=V. For example, X←5 ¯3 8 4 is lean while Y←5 ¯3 5 4 is not.

```
      X                              Y
5 ¯3 8 4                       5 ¯3 5 4

      +/X∘.=X                        +/Y∘.=Y
1 1 1 1                        2 1 2 1
```

We need only consider lean vectors of arguments for fitting polynomials. For example, consider the vector of arguments X←2 3 2 4 and suppose that 10 7 10 ¯4 is the vector of values G X. Evidently these vectors contain redundant information; the same information is contained in the vectors 1 1 0 1/X and 1 1 0 1/G X, and moreover 1 1 0 1/X is lean.

If X is a lean vector then the matrix (⍴X) VAN X is a valid right argument of domino. For example:

```
      0≠DET 4 VAN 0 ¯3 1 5
1                                                    ⌹54
```

Thus if X is lean the solution C of Equation 3.7.2 is identical to (G X)⌹(⍴X) VAN X. This suggests the following definition:

```
      CFP:ω⌹(⍴α) VAN α
```

(for the coefficient vector of a fitting polynomial); if Y is the vector of values of a function G for a vector of arguments X, then ∧/Y=(X CFP Y) POLY X and the polynomial with coefficient vector X CFP Y is a fitting polynomial for G. For example, if G is the square root function G:ω*0.5:

```
      X←1 4 9 16
      C←X CFP G X
      G X                            C POLY X
1 2 3 4                        1 2 3 4               ⌹57
```

It is also possible to produce fitting polynomials centered other than at 0. Specifically, for a scalar A, a lean vector X, and the vector CA←(X-A) CFP G X the values of the polynomial CA POLY ω-A and the function G ω agree for the vector of arguments X.

4

Applications of the Derivative

4.1 INTRODUCTION

Up to now we have been mainly concerned with polynomials, but as we look more closely at applications of the derivative it is helpful to also look beyond polynomials. For that purpose we will use the power function *POW*:α⋆ω. If *B* is a non-negative integer the function *F*:α⋆*B* is a polynomial, but otherwise *F* is not a polynomial.

For a non-yegative integer *B* the function 1value *A POW B* can be defined in terms of the times function. Specifically:

```
A POW B                                        [4.1.1]
×/BρA
```

Using this identity it is a simple matter to show for non-negative integers *B* and *C* that

```
A POW B+C                A POW B×C             [4.1.2]
(A POW B)×A POW C        (A POW B) POW C
```

For right arguments that are not non-negative integers the values of *POW* are defined implicitly so that the Identities 4.1.2 remain valid. For example, if *A* is non-negative then *A POW* 0.5 is defined to be the unique non-negative scalar *R* for which *A*=*R*×*R*. The reason for this can be seen from the left side of 4.1.2:

```
A
A POW 1
A POW 0.5+0.5
(A POW 0.5)×A POW 0.5
R×R
```

This definition of the power function is implicit because it does not provide a procedure for producing values of *POW*. A few values are obvious, but most are not. For example, 4 *POW* 0.5 is obviously 2 (or ¯2). At several points in this chapter we will be concerned with producing values of *POW*. Explicit definitions of *POW* will also be discussed in Chapter 7.

For an integer *R* the derivative of *F*:α *POW R* is *DF*:*R*×α *POW R*-0≠*R*. In fact this is true for every *R*, not just integers. For example, the derivative of *G*:α *POW* 1.7 is *DG*:1.7×α *POW* 0.7 and the derivative of *H*:α *POW* ¯1.7 is

D̲H:¯1.7×α POW ¯2.7. This derivative rule will be used throughout this chapter, although its proof is deferred to Chapter 7. ⌹1

Note that the derivative D̲F of F:ω POW R can be defined in terms of F, and consequently if a value F X is known then the value D̲F X is also known. To see how D̲F is defined in terms of F, consider the sequence of identities:

```
D̲F ω
R×ω POW R-0≠R
R×(ω POW R)×ω POW -0≠R
(F ω)×R×ω POW -0≠R
```

It follows from the left side of 4.1.2 that ω POW -0≠R is equivalent to ÷ω⋆0≠R, and consequently D̲F can be defined as D̲F:(F ω)×R÷ω⋆0≠R. This definition of D̲F will be used at various times in this chapter to stress the fact that the value D̲F X is known whenever the value F X is known. Note, however, that this definition has the disadvantage that its value at zero is indeterminate when 0≠R.

4.2 APPLICATIONS OF THE DERIVATIVE

<u>Critical Arguments</u>. The argument A of a function F is called a <u>critical argument</u> and the coordinate point A,F A is called a <u>critical point</u> of the graph of F if 0=D̲F A.

Geometrically, A is a critical argument of F if the line tangent to the graph of F at the point A,F A has slope 0. We distinguish three types, all of which are illustrated by the polynomial P:1 0 0 ¯5 0 3 POLY ω whose derivative is D̲P:0 0 ¯15 0 15 POLY ω, or equivalently 15××/ω∘.-0 0 ¯1 1. The critical arguments of P are 0, ¯1, and 1. The graph of P is illustrated in Figure 4.2.1. The value P ¯1 is called a <u>local maximum</u> of P because it is larger than all other values P X for arguments X near ¯1. Analogously, the value P 1 is called a <u>local minimum</u> of P. The critical point 0,P 0 is called an <u>inflection point</u>.

The concepts of local maximum and local minimum can be defined without reference to derivatives. The argument A is a <u>local maximum</u> of F if F A is greater than or equal to F X for all arguments X near A. In other words, there is a small, positive scalar E for which

```
(E>|X-A)≤(F A)≥F X
```

Analogously, A is a <u>local minimum</u> of F if there is a small positive scalar E for which

```
(E>|X-A)≤(F A)≤F X
```

Inflection points will be discussed in Section 4.4. ⌹4

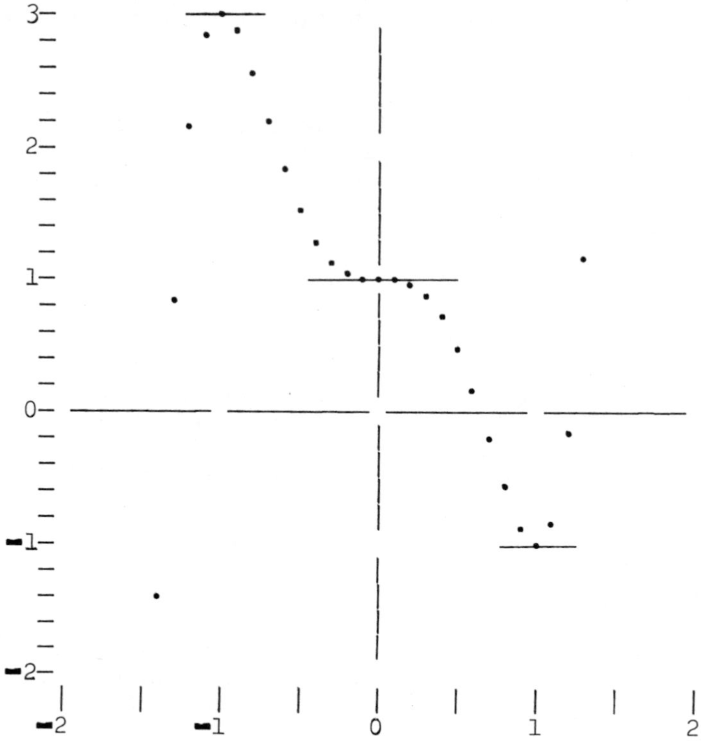

TYPES OF CRITICAL ARGUMENTS
FIGURE 4.2.1

Monotone Functions. The function *F* is said to be increasing on the interval (*A*≤ω)∧ω≤*B* if for any pair of arguments in the interval, the sign of the difference of the values of *F* equals the sign of the difference of the arguments. Formally:

```
(×-/X,Y)=×-/F X,Y
```

Equivalently, *F* is increasing if every secant line joining pairs of points on the graph of *F* has positive slope. Formally:

```
(X≠Y)≤0<((F X)-F Y)÷X-Y
```

Analogously, *F* is said to be decreasing on the interval (*A*≤ω)∧ω≤*B* if

```
(×-/X,Y)=-×-/F X,Y
```

or

```
(X≠Y)≤0>((F X)-F Y)÷X-Y
```

F is said to be strictly monotonic on the interval (*A*≤ω)∧ω≤*B* if *F* is either increasing or decreasing on this interval. For example, the polynomial *P*:0 ¯3 0 1 *POLY* ω is increasing of the intervals ω≤¯1 and ω≥1 and decreasing on the interval (¯1≤ω)∧ω≤1, and is therefore strictly monotonic on each of the three intervals, as illustrated in Figure 4.2.2.

As another example, consider the function *G*:ω*0.5, or equivalently ω *POW* 0.5, where *POW* is the function discussed in Section 4.1. The derivative of *G* ω is *DG*:0.5÷*G* ω. Since 0<*G* ω on the interval 0<ω, it follows that 0<*DG* ω on this interval. Consequently *G* ω is an increasing function. Since 0=*G* 0, it follows that 0 is not in the domain of *DG*. However, for arguments *X* near 0, the values *G X* are also close to 0, and therefore the values *DG X* are large. Therefore, for arguments *X* near 0, the tangent lines to the graph of *G* at the points *X*,*G X* are steep. Moreover these tangent lines approach the horizontal axis as *X* is chosen closer and closer to 0. The graph of *G* is shown in Figure 4.2.3.

If 0<*DF* ω for all arguments in the interval (*A*≤ω)∧ω≤*B*, then obviously *F* is increasing on this interval. Similarly, if 0>*DF* ω then *F* is decreasing. The converse statements are not in general true; it is possible that *F* is monotonic on

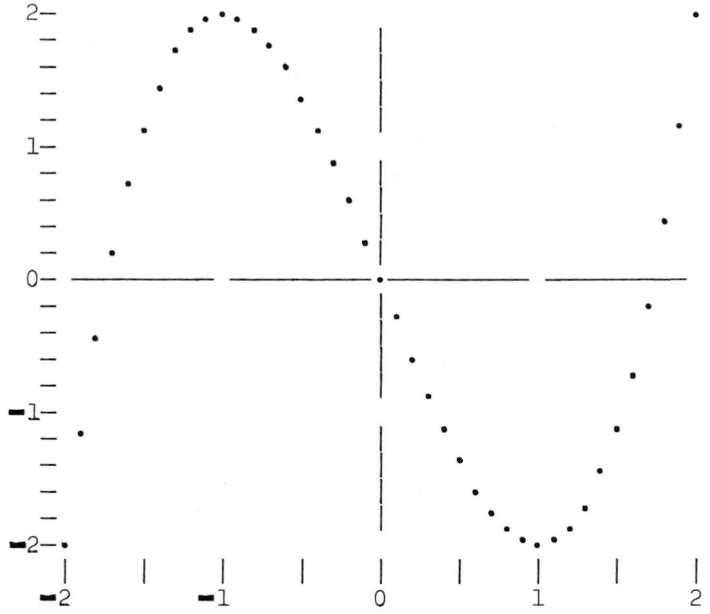

GRAPH OF 0 ¯3 0 1 *POLY* ω
FIGURE 4.2.2

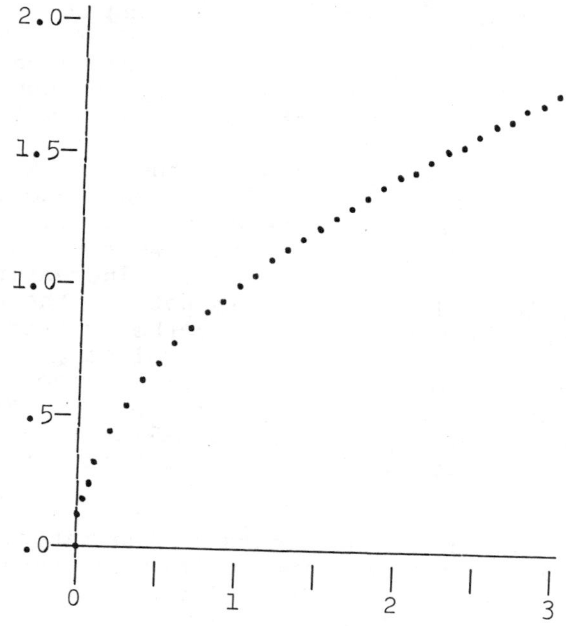

GRAPH OF $\omega\star0.5$
FIGURE 4.2.3

an interval and that $0=\underline{D}F\ X$ for a few arguments X in the interval. For example, consider the function $F{:}\omega\star3$ and its derivative $\underline{D}F{:}3\times\omega\star2$. F is increasing on every interval and in particular on the interval $2\geq|\omega$. However, $0=\underline{D}F\ 0$.

The idea of strict monotonicity can be broadened to allow for functions with horizontal segments in their graphs. The function F is said to be <u>non-decreasing</u> on the interval $(A\leq\omega)\wedge\omega\leq B$ if every secant line joining pairs of points on the graph of F has non-negative slope. <u>Non-increasing</u> is defined analogously.

F is said to be <u>monotonic</u> on the interval $(A\leq\omega)\wedge\omega\leq B$ if F is either non-decreasing or non-increasing on the interval. Monotonic functions that are not strictly monotonic will arise in the next section. The simplest examples are those whose graphs are horizontal straight lines.

<u>Tangent line Approximations</u>. The function

$$TF{:}(F\ \alpha)+(\underline{D}F\ \alpha)\times\omega-\alpha$$

is called the <u>tangent line function</u> of F. The graph of the function $A\ TF\ \omega$ is the line tangent to the graph of F at the coordinate point $A,F\ A$.

Certain values of the tangent line function *TF* are rough approximations to values of *F*. Specifically, for a small value of the scalar *S* the function value *F A+S* is approximately *A TF A+S* (Figure 4.2.4). For example, if *G*:ω⋆0.5 then *DG*:(*G* ω)×0.5÷ω. Evidently *G* 4 is 2 and *DG* 4 is 0.25. Therefore the function 4 *TG* ω is equivalent to *T*4:2+0.25×ω-4. The graph of *T*4 is the line tangent to the graph of *G* at the point 4 2. The function value *G* 4.01 is, approximately, *T*4 4.01. Moreover, it is easy to produce the value *T*4 4.01. Namely, *T*4 4.01 is equal to 2+0.25×4.01-4, or 2.0025. ⌹30

Note that the tangent line function can be defined as

TF:((*F* α),*DF* α) *POLY* ω-α [4.2.5]

A Rootfinder. The argument *R* is called a root or zero of *F* if 0=*F R*. To produce the roots of *F* we must solve the equation 0=*F* ω. While it is not always possible to produce exact solutions of this equation there are many procedures for producing approximate solutions; the one presented here is called Newton's method.

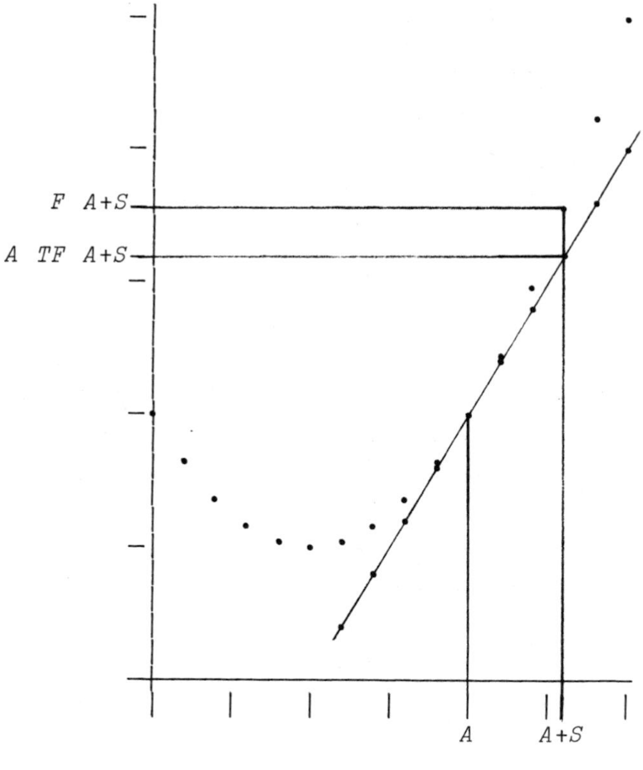

TANGENT LINE APPROXIMATION
FIGURE 4.2.4

Newton's method can be described geometrically as follows. Suppose that $X0$ is an approximation to a solution X of the equation $0{=}F\ \omega$. Sketch the line tangent to the graph of F at coordinate point $X0,F\ X0$. If this line is not horizontal then it intersects the horizontal axis in one and only one point. Let $X1,0$ be the coordinates of the intersection point. For the function F whose graph is illustrated in Figure 4.2.6, $X1$ is clearly a better approximation to X than is $X0$. Now repeat the procedure with $X1$ in place of $X0$. That is, sketch the graph of the line tangent to the graph of F at the coordinate point $X1,F\ X1$ and let $X2,0$ be the coordinates of the intersection of the tangent line and the horizontal axis. As before, in Figure 4.2.6, $X2$ is clearly a better approximation to X than is $X1$. Evidently, by continuing this procedure we obtain better and better approximations to a solution X of the equation $0{=}F\ \omega$.

In order for Newton's method to become an effective tool we must rephrase the above geometric description in terms of functions. For that purpose we will use the tangent function $TF{:}(F\ \alpha){+}(\underline{D}F\ \alpha){\times}\omega{-}\alpha$. If A is an approximation to the solution X of the equation $0{=}F\ \omega$ and AN is the next approximate produced by Newton's method, then AN is the solution of the equation $0{=}A\ TF\ \omega$. The following equations are equivalent:

$$0{=}A\ TF\ \omega$$
$$0{=}(F\ A){+}(\underline{D}F\ A){\times}\omega{-}A$$
$$(\omega{-}A){=}{-}(F\ A){\div}\underline{D}F\ A$$
$$\omega{=}A{-}(F\ A){\div}\underline{D}F\ A$$

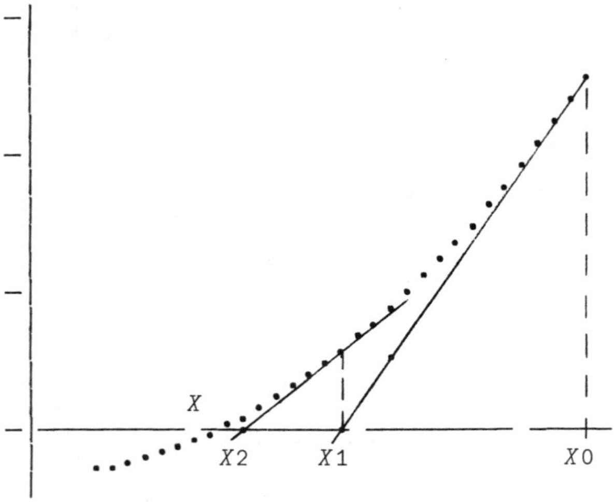

NEWTON'S METHOD
FIGURE 4.2.6

Using the last equation as a guide we define the tangent line intersection function

```
ITF:α-(F α)÷D̲F α
```

so that the last equation is equivalent to ω=ITF A.

For example, for the function F:4-ω*2 a solution of the equation 0=F ω is, formally, 2*0.5. We can produce approximations to this solution by using the tangent intersection function ITF:α-(F α)÷D̲F α and the initial approximation 2. Note that D̲F:2×ω.

```
      AN←ITF 2
      AN
1.5
      AN←ITF AN
      AN
1.416666667

      AN←ITF AN
      AN
1.414215686

      AN←ITF AN
      AN
1.414213562
```

Newton's method is a recursive procedure that produces approximations to a solution of the equation 0=F ω by improving each previous approximation. The method is obviously unending. In order for the method to be of practical value, the process must be stopped after a finite number of steps. Suppose we agree to accept an approximation which gives, say, at least six decimal place accuracy. An accepted rule of thumb is to stop the procedure at a point when the first six decimal digits of an approximation A agree with the first six decimal digits of the next approximation AN. That is, the iteration is stopped when 0.5E¯6>|A-AN, so that all digits in 6⍕|A-AN are 0. This strategy is called a stopping rule. ⌹31

Using the preceding discussion as a guide, we define the r̲oot function of F by

```
R̲F:R̲F ITF ω : 0.5E¯6>⌈/|ω-ITF ω : ITF ω
```

This function produces approximate values of the roots of F. The argument is an initial, rough approximation and is sometimes called a starting value.

For example, the roots of F:4.01-ω*2 are, formally, 4.01*0.5 and ¯4.01*0.5. Approximate values of either of these roots, or both, are produced as follows (using the initial approximations 2 and ¯2):

```
      R̲F 2
2.002498439

      R̲F ¯2 2
¯2.002498439 2.002498439
```

Recall that the value 4.01*0.5 was previously estimated to be 2.0025 using the tangent line approximation. In general, for the function

```
      ITP:ω-(α POLY ω)÷(PDERIV α) POLY ω
```

the coordinate point (*C ITP A*),0 is the intersection point of the line tangent to the graph of the polynomial *C POLY ω* at *A*,*C POLY A* and the horizontal axis. Also, for the function

```
      R̲POLY:α R̲POLY α ITP ω : 0.5E¯6>⌈/|ω-α ITP ω : α ITP ω
```

if *A* is a sufficiently good approximation to a root of the polynomial *C POLY ω* then *C R̲POLY A* is a six decimal digit approximation to this root. ⌹32

4.3 FUNCTIONS OF BOUNDED VARIATION

Many functions of practical interest are characterized by the fact that every interval in their domain can be subdivided in such a way that the function is monotonic on each piece of the subdivision. For example, polynomials have this property. Specifically, an appropriate subdivision of the interval 2≥|ω for the polynomial *P*:0 ¯3 0 1 *POLY ω* consists of the intervals with endpoints ¯2 and ¯1, ¯1 and 1, and 1 and 2 (Figure 4.2.2). Every function so characterized has the property that it can be expressed as the sum of two monotonic functions, one non-increasing and one non-decreasing, on each interval in its domain. This property, which is important because it allows us to analyze functions in terms of two other functions with relatively simple graphs, will be used in the next chapter. The following example shows how to construct the two monotonic functions.

Consider the polynomial *P*:1 ¯6 5.5 ¯2 0.25 *POLY ω* on the interval (0≤ω)∧ω≤4, whose graph is Figure 4.3.1. If

```
      V←0 1 2 3 4
```

then *P* is either increasing or decreasing on each interval (*V*[*I*]≤ω)∧ω≤*V*[*I*+1] ; the latter will be called the *I*th <u>subinterval</u> <u>of</u> *V*. If

```
      IND:¯1++/ω∘.≥¯1↓α
```

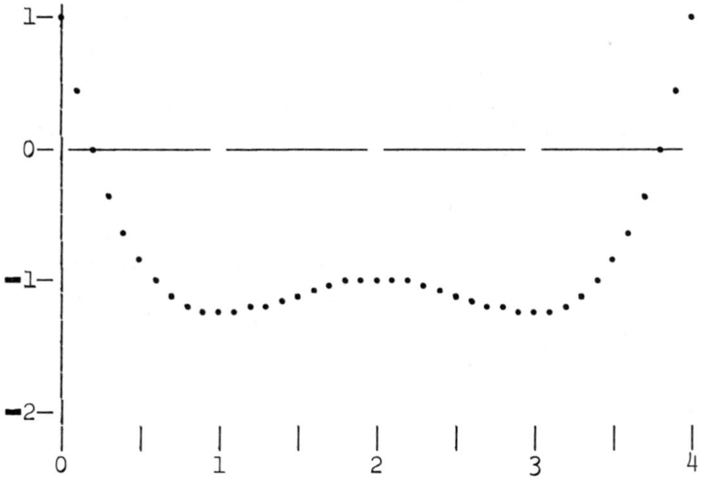

GRAPH OF *P*:1 ¯6 5.5 ¯2 0.25 *POLY* ω
FIGURE 4.3.1

then *V IND X* is the index of the (rightmost) subinterval of *V* that contains *X*. For example:

```
      V←0 1 2 3 4
      V IND 0 2.5
0 2
```

If

```
      RISE:SUM (0≤DIFF α)×DIFF α
```

then the *I*th element of *RISE P V* is the total increase in the values of *P* from *V*[0] to *V*[*I*], neglecting all decreases in values. Continuing the above example:

```
      P V
1 ¯1.25 ¯1 ¯1.25 1

      RISE P V
0 0 0.25 0.25 2.5
```

Thus if

```
      RSP:(RISE P α)[α IND ω]
```

then the value *V RSP X* is the total increase in values of *P* from *V*[0] to the left endpoint of the subinterval of *V* that contains *X*. For example:

```
      V RSP 2.87
0.25
```

Analogously, we have

```
FALL:SUM (0>DIFF α)×DIFF α
```

and the *I*th element of *FALL P V* is the total decrease in the values of *P* from *V*[0] to *V*[*I*], neglecting all increases in values. Also:

```
FLP:(FALL P α)[α IND ω]
```

and the value *V FLP X* is the total decrease in the values of *P* from *V*[0] to the left endpoint of the subinterval of *V* that contains *X*.

```
      P V
1 ¯1.25 ¯1 ¯1.25 1

      FALL P V
0 ¯2.25 ¯2.25 ¯2.5 ¯2.5

      V FLP 2.5
¯2.25
```

Next, we define

```
INCP:(P ω)-P α[α IND ω]
```

The value *V INCP X* is the difference (or increment) between *P X* and the value of *P* at the left endpoint of the subinterval of *V* that contains *X*. Finally, we define

```
RP:(α RSP ω)+(0≤α INCP ω)×α INCP ω          [4.3.2]
```

and

```
FP:(α FLP ω)+(0>α INCP ω)×α INCP ω          [4.3.3]
```

The graphs of *V RP ω* and *V FP ω* are illustrated in Figures 4.3.4 and 4.3.5 respectively. Note that the rising segments of the graph of *V RP ω* are at most vertical translations of the rising segments of the graph of *P*, and similarly for the falling segments of *V FP ω* and *P*. Also, the horizontal segments of *RP* and *FP* occur over the subintervals where *P* is non-increasing and non-decreasing respectively. It is easy to check graphically that *P* is the sum of the functions (*P V*[0])+*V RP ω*, which is non-decreasing, and *V FP ω*, which is non-increasing. For example:

```
      X←0.2×ι21

      ∧/(P X)=(P V[0])+(V RP X)+V FP X
1
```

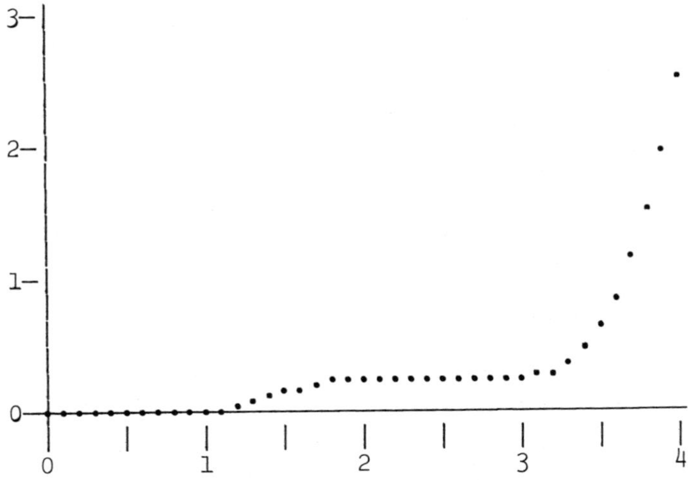

THE RISING COMPONENT OF 0 ¯3 0 1 *POLY ω*
FIGURE 4.3.4

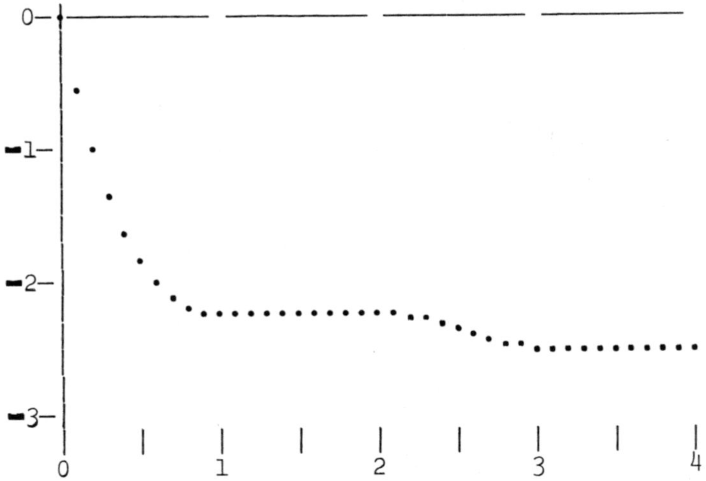

THE FALLING COMPONENT OF 0 ¯3 0 1 *POLY ω*
FIGURE 4.3.5

The graphs of the functions *V FP ω* and *V RP ω* in the above example can be produced directly from the graph of *P* and the vector

```
      W←×DIFF P V
      W
¯1 1 ¯1 1
```

If $W[I]$ is 1, then the graph of V RP ω on the Ith interval of V is a vertical translation of the graph of P, while the graph of V FP ω is a horizontal line segment. If $W[I]$ is ¯1, then the graph of V FP ω is a vertical translation of the graph of P, while the graph of V RP ω is a horizontal line segment. Evidently, if $W[I]$ is 0, the graphs of both V FP ω and V RP ω on the Ith interval of V are horizontal line segments.

We now know the shapes of the graphs of V FP ω and V RP ω on the subintervals of V. It remains to determine the appropriate vertical translations of these segments that produce the graphs of V FP ω and V RP ω. The vertical translations are chosen so that both V FP $V[0]$ and V RP $V[0]$ are 0, and so there are no breaks in the completed graphs of V FP ω and V RP ω.

Thus to graph V FP ω and V RP ω we can first sketch their graphs on the zeroth subinterval of V, aligning the segments so that both V FP $V[0]$ and V RP $V[0]$ are 0. Next, sketch the graph of V RP ω on the first subinterval of V, aligning this segment so that it joins the first segment at the point over $V[1]$. Do the same thing for V FP ω, then repeat the procedure on the second subinterval of V, and so on.   ⊟40

A function that can be represented as the sum of two monotonic functions on each interval $(A \le \omega) \wedge \omega \le B$ in its domain is said to be of <u>bounded variation.</u> While it is often difficult and sometimes impossible to produce a subdivision V for which a function is monotonic on every subinterval of V, it is still true that almost all monadic scalar functions of practical interest can be expressed as the sum of two monotonic functions. The construction of these monotonic functions is similar to that above and will be discussed in the exercises.   ⊟42

4.4   APPLICATIONS OF THE SECOND DERIVATIVE

For a function F and its derivative $\underline{D}F$, the derivative of $\underline{D}F$ is denoted $\underline{DD}F$ and is called the <u>second derivative</u> of F. For example, if $F:\omega*4$ then $\underline{D}F:4\times\omega*3$ and $\underline{DD}F:12\times\omega*2$. More generally, if $F:\omega*R$ and $0 \ne R$ then $\underline{D}F:R\times\omega*R-1$ and $\underline{DD}F:R\times(R-1)\times\omega*R-2$, or, if $\omega \ne 0$, $\underline{D}F:(F\ \omega)\times R \div \omega$ and $\underline{DD}F:(F\ \omega)\times R\times(R-1)\div\omega*2$.

<u>Flection</u> <u>and</u> <u>Concavity.</u> Since $\underline{DD}F$ is the derivative of $\underline{D}F$, $\underline{D}F$ is an increasing function on an interval $(A \le \omega)\wedge\omega \le B$ if $0 < \underline{DD}F$ ω for all arguments in the interval. Also, $\underline{D}F$ is a decreasing function on this interval if $0 > \underline{DD}F$ ω for all arguments. The geometric meaning of an increasing or decreasing derivative can be seen from the following example. If $F:0$ ¯3 0 1 $POLY$ ω the $\underline{D}F:$¯3 0 3 $POLY$ ω and $\underline{DD}F:0$ 6 $POLY$ ω. Since the values of $\underline{DD}F$ are negative for

negative arguments, $\underline{D}F$ is decreasing on the interval $0>\omega$; also, the values of $\underline{D}\underline{D}F$ are positive for positive arguments and so $\underline{D}F$ is increasing on the interval $0<\omega$ (Figure 4.2.2).

Place a ruler on the graph of F at the coordinate point $^{-}2,^{-}2$ in such a way that the edge of the ruler is tangent to the graph; the edge of the ruler represents the line tangent to the graph of F at the coordinate point $^{-}2,^{-}2$. Imagine that the argument X starts at $^{-}2$ and slowly increases so that the coordinate point $X,F\ X$ moves slowly to the right along the graph of F. As this coordinate point moves along the graph of F, move the ruler so that its edge continually represents the line tangent to the graph of F at the moving coordinate point $X,F\ X$. As the argument X increases from $^{-}2$ to 0 the ruler turns in a clockwise direction, indicating that the derivative $\underline{D}F$ is decreasing on the interval $(^{-}2\le\omega)\wedge\omega\le 0$. As the argument X increases from 0 the ruler turns in a counterclockwise direction, indicating that $\underline{D}F$ is increasing on the interval $0\le\omega$.

Note that for each argument X for which $\underline{D}\underline{D}F\ X$ is negative the edge of the ruler lies above a segment of the graph near the coordinate point $X,F\ X$, indicating that the graph bends downward near that point. Also, for each argument X for which $\underline{D}\underline{D}F\ X$ is positive the edge of the ruler lies below a segment of the graph near the coordinate point $X,F\ X$, indicating that the graph bends upward near this point. Finally, note that the larger the value of $|\underline{D}\underline{D}F\ \omega$ the faster the ruler turns, indicating the greater the bending of the graph.

As a second example, consider the function $G{:}\omega\star 0.5$, whose second derivative is $\underline{D}\underline{D}G{:}{}^{-}0.25\times\omega\star{}^{-}1.5$. Since the values of G are positive on the interval $0<\omega$, the values of $\underline{D}\underline{D}G$ are negative, and this indicates that the graph of G bends downward on the interval $0<\omega$ (Figure 4.2.3).

For a function F and its second derivative $\underline{D}\underline{D}F$, the function value $\underline{D}\underline{D}F\ X$ is called the bending or flection of F at X. F is said to be concave downward on the interval $(A\le\omega)\wedge\omega\le B$ if $0>\underline{D}\underline{D}F\ \omega$ for all arguments in this interval, and F is said to be concave upward on this interval if $0<\underline{D}\underline{D}F\ \omega$ for all arguments in the interval. For instance, in the above example F is concave downward on the interval $0>\omega$ and concave upward on the interval $0<\omega$ (Figure 4.2.2). In that example the coordinate point $0,0$ on the graph of F is called an inflection point. In general, the coordinate point $C,F\ C$ on the graph of F is called an inflection point if a segment of the graph to one side of the point bends downward and another segment of the graph to the other side of the point bends upward. In most cases, $0=\underline{D}\underline{D}F\ C$ if the coordinate point $C,F\ C$ is an inflection point.

Concavity, like monotonicity, can be defined without appealing to derivatives. For example, suppose that the function `F` is concave downward on the interval `(A≤ω)∧ω≤B`. Then for every pair of arguments `C` and `D` in this interval, the graph of `F` lies above the segment of the secant line joining the coordinate points `C,F C` and `D,F D`. In particular, let `S←0.5×D-C` so that the mid-point of this segment has coordinates defined by `(C+S),0.5×(F C)+F C+2×S` (Figure 4.4.1). Then

```
(F C+S)>0.5×(F C)+F C+2×S
```

or equivalently

```
0>1 ¯2 1 +.×F C+0,S,2×S
```

or

```
0>DIFF DIFF F C+S×⍳3
```

Thus we can define `F` to be concave downward on the interval `(A≤ω)∧ω≤B` if for scalars `C` and `S` such that `0≠S` and `∧/(A<C+S×⍳3),B>C+S×⍳3`:

```
0>DIFF DIFF F C+S×⍳3
```

Similarly, we can define `F` to be concave upward on this interval if for scalars `C` and `S` as above:

```
0<DIFF DIFF F C+S×⍳3
```

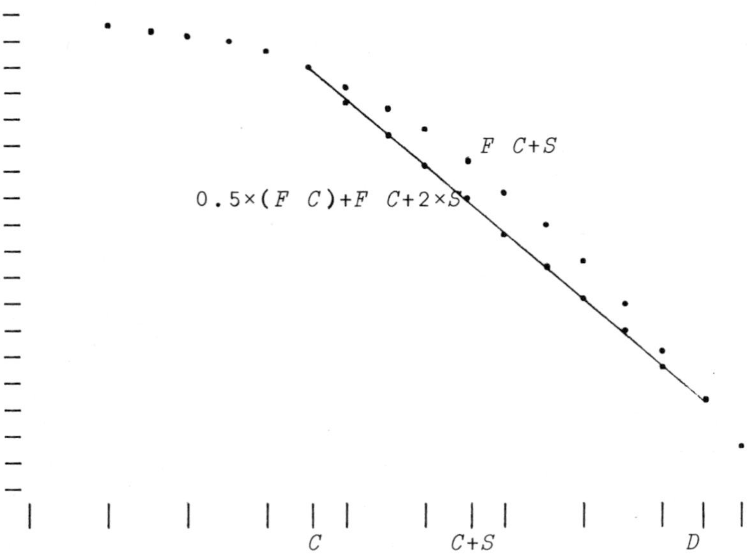

SECANT LINES AND CONCAVITY
FIGURE 4.4.1

Quadratic Approximations. For a function *F*, its tangent line function *TF*:(*F* α)+(D̲*F* α)×ω-α, and an argument *A* the function *A TF* ω is a rough approximation to *F* for arguments near *A*. The value and slope of *A TF* ω at *A* are equal to those of *F* at *A*; however, the flection of *F* at *A* is not necessarily equal to the flection of *A TF* ω at *A*, which is zero. Thus the graphs of *F* and *A TF* ω do not necessarily bend the same way at the point *A*,*F A*. We should suspect that if the value, slope, and flection of a function at *A* are equal to those of *F* at *A*, then that function is a better approximation to *F* than is *A TF* ω for arguments near *A*. In fact, this is in general true. One such function is *A QF* ω, where *QF* is the quadratic function

    QF:(α TF ω)+0.5×(D̲D̲F α)×(ω-α)*2

For example, if *G*:ω*0.5 then D̲*G*:(*G* ω)*0.5÷ω and D̲D̲*G*:(*G* ω)*¯0.25÷ω*2. Also, *G* 4 is 2, D̲*G* 4 is 0.25 and D̲D̲*G* 4 is ¯.03615. Thus the function 4 *QG* ω is equivalent to *Q*4:2+(0.25×ω-4)+¯.018075×(ω-4)*2. It is easy to check that *Q*4 4 is equal to *G* 4, D̲*Q*4 4 is equal to D̲*G* 4, and D̲D̲*Q*4 4 is equal to D̲D̲*G* 4. The function value *G* 4.01 is approximately *Q*4 4.01, which is 2.002498437. According to the last example of Section 4.2, this approximation is accurate to at least 6 decimal digits. The corresponding approximation produced by the tangent function is 2.0025, which evidently is not as accurate.

Note that the quadratic function *QF* can be defined by

    QF:((F α),(D̲F α),0.5×D̲D̲F α) POLY ω-α   [4.4.3]

⌹50

4.5 TAYLOR POLYNOMIALS

We have seen that the tangent line function *TF*:((*F* α),D̲*F* α) *POLY* ω-α approximates *F* and the quadratic function *QF*:(α *TF* ω)+(0.5×D̲D̲*F* α)×(ω-α)*2 approximates it even better. We should suspect that there is a cubic function that produces still better approximations, a quartic function that is even better than the cubic function, and so on. In fact there are such functions, and they are called the Taylor functions of *F* (for the eighteenth century English mathematician B. Taylor).

Recall that the derivative of the derivative of *F* is called the second derivative of *F*; in general, the derivative of second derivative is called the third derivative, the derivative of the third derivative is called the fourth derivative, and so on. In this regard the derivative of *F* is called the first derivative of *F*, and *F* is called the zeroth derivative of itself. In this section we will use the names D̲0*F* for the zeroth derivative, D̲1*F* for the first derivative, D̲2*F* for the second derivative, and so on.

The Taylor functions of F are defined as follows:

The first order Taylor function is

```
T1F:(F α)+(DF α)×ω-α
```

That is, $T1F$ is the tangent function of F. $T1F$ is called the first order Taylor function of F because for each scalar A, the function $A\ T1F\ \omega$ is a polynomial of order 1, called the first order Taylor polynomial of F at A;

The second order Taylor function of F is

```
T2F:(α T1F ω)+((÷!2)×D2F α)×(ω-α)*2
```

That is, $T2F$ is the quadratic function of F, and for each scalar A the function $A\ T2F\ \omega$ is a polynomial of order 2, called the second order Taylor polynomial of F at A;

The third order Taylor function of F is

```
T3F:(α T2F ω)+((÷!3)×D3F α)×(ω-α)*3
```

For each scalar A, the function $A\ T3F\ \omega$ is a polynomial of order 3, called the third order Taylor polynomial of F at A;

The fourth order Taylor function of F is

```
T4F:(α T3F ω)+((÷!4)×D4F α)×(ω-α)*4
```

and so on;

In general, if $\alpha\ B\ \omega$ is the (N-1)th order Taylor function of F and $D\alpha$ is the Nth derivative of F, then the Nth order Taylor function of F is

```
C:(α B ω)+(÷!N)×(D α)×(ω-α)*N
```

Also, the function $A\ C\ \omega$ is a polynomial of order N called the Nth order Taylor polynomial of F at A.

Note that, according to the definition of $A\ C\ \omega$, the coefficient of the term $(\omega\text{-}A)\star I$ in the Nth order Taylor polynomial of F at A is ÷!I times the value of the Ith derivative of F at A.

As an example, consider the function $\omega\star R$ for a non-zero scalar R. Table 4.5.1 is a list of the derivatives of this function. In particular, the Taylor polynomials of $\omega\star 0.5$ at 4 are

```
T1:2+(0.25)×ω-4
T2:(T1 ω)+(÷!2)×¯0.03125×(ω-4)*2
T3:(T2 ω)+(÷!3)×0.01171875×(ω-4)*3
```

and so on. Table 4.5.2 provides a list of values of these

Taylor polynomials for the argument 4.01. Note that the higher the order of the polynomial the better the approximation to 4.01⋆0.5, whose true value to 10 decimal digits is 2.0024984395.

DERIVATIVE	EXPRESSION
0	ω⋆R
1	(ω⋆R)×R÷ω
2	(ω⋆R)×R×(R-1)÷ω⋆2
3	(ω⋆R)×R×(R-1)×(R-2)÷ω⋆3
N	(ω⋆R)×(×/R-⍳N)÷ω⋆N

DERIVATIVES OF ω⋆R
TABLE 4.5.1

ORDER	VALUE
1	2.0025
2	2.0024984375
3	2.0024984395

TAYLOR POLYNOMIALS OF ω⋆0.5 AT 4 EVALUATED FOR 4.01

TABLE 4.5.2

In general, according To Table 4.5.1, element *I* of the vector

```
(ω⋆R)×(1,×\R-⍳α)÷ω⋆⍳α+1
```

is the *I*th derivative of ω⋆R. Thus the value *N* T̲F *A* of the function

```
T̲F:(ω⋆R)×(1,×\R-⍳α)÷(ω⋆⍳α+1)×!⍳α+1                [4.5.3]
```

is the coefficient vector of the *N*th order Taylor polynomial of *F*:ω⋆R at *A*. The function T̲F is called the Taylor coefficient function of *F*. ⊞52

<u>Taylor Coefficient Functions</u>. To each function *H* there is an associated function T̲H, called the <u>Taylor coefficient function of</u> *H*, whose value *N* T̲H *A* is the coefficient vector of the *N*th order Taylor polynomial of *H* at *A*. Thus the *N*th order Taylor function of *H* is

```
(N T̲H α) POLY ω-α
```

The importance of Taylor polynomials lies in the approximation and construction of non-polynomial functions. Their use in constructing new functions will be studied in detail in Chapter 6. In order that these polynomials can be effective tools for approximation, we must be able to produce their coefficient vectors. That is, we must be able to produce explicit definitions of Taylor coefficient functions. Evidently, the Taylor coefficient function $\underline{T}H$ can be defined in terms of H, $\underline{D}H$, $\underline{DD}H$, and so on, using the definition of Taylor polynomials. However, this is not a practical definition of $\underline{T}H$, since it requires all higher derivatives of H to be defined.

For certain functions there is a single expression that produces values of the function and all its higher derivatives. For example, such an expression for the function *ω⋆R* can be found above. As illustrated there, this expression can serve as the basis of a definition of the associated Taylor coefficient function.

In general, it is difficult to construct a single expression for a function H that produces values of H and all its higher derivatives, and therefore it is difficult to define its Taylor coefficient function. However, we can develop an algebra of Taylor functions yielding rules similar to those developed for cline functions and derivatives, and used in a similar way.

For this purpose we will adopt the notation $\underline{T}F$ for the Taylor function of F, and the functions *PLUS*, *TIMES*, etc. from Chapter 3. Specifically, we have the following rules for Taylor coefficient functions:

<u>Plus Rule for Taylor Coefficient Functions.</u> If *P ω* is equivalent to *(F ω)+G ω*, and if $\underline{T}P$, $\underline{T}F$, and $\underline{T}G$ are the Taylor coefficient functions of *P*,*F*, and *G* respectively, then

α $\underline{T}P$ ω
(α $\underline{T}F$ ω) *PLUS* α $\underline{T}G$ ω

<u>Times Rule for Taylor Coefficient Functions.</u> If *T ω* is equivalent to *(F ω)×G ω*, then

α $\underline{T}T$ ω
(α+1)↑(α $\underline{T}F$ ω) *TIMES* α $\underline{T}G$ ω

<u>Composition Rule for Taylor Coefficient Functions.</u> If *C ω* is equivalent to *F G ω*, then

α $\underline{T}C$ ω
(α $\underline{T}F$ *G* ω) *COMP* 0,1↓α $\underline{T}G$ ω

<u>Reciprocal Rule for Taylor Coefficient Functions.</u> If *R ω* is equivalent to *÷F ω*, then

α $\underline{T}R$ ω
(α+1) *REC* α $\underline{T}F$ ω

<u>Taylor Coefficient Function Rule for Derivatives.</u> For the Taylor coefficient function $\underline{TD}H$ of the derivative $\underline{D}H$ of H we

have:

α T̲D̲H ω
DERIV (α+1) T̲H ω

<u>Taylor Coefficient Function Rule for Integrals</u>. If *H* is an integral of *F* then:

α T̲H ω
(H ω) INTEG (α-1) T̲F ω

when α≠0, and otherwise 0 *T̲H ω* is 1ρ*H ω*. That is, *T̲H* can be defined as

T̲H:(H ω) INTEG (α-1)T̲F ω : 0=α : 1ρH ω

The basis of these rules can be understood by comparing them with the algebra of polynomials described in Section 3.6. For example, if *B* and *C* are coefficient vectors of the *N*th order Taylor polynomials at *A* of the functions *F ω* and *G ω* respectively, then the values of *B POLY ω-A* and *C POLY ω-A* are approximately equal to the values of *F ω* and *G ω* for arguments near *A*. Consequently the values of *(B PSUM C) POLY ω-A* and *(B PPROD C) POLY ω-A* are approximately equal to the values *(F ω)+G ω* and *(F ω)×G ω* for arguments near *A*. Thus it is reasonable to expect that *(B PSUM C) POLY ω-A* and *(B PPROD C) POLY ω-A* are closely related to Taylor polynomials of *(F ω)+G ω* and *(F ω)×G ω*.

The proofs of these rules will be illustrated with a sketch of the proof of the times rule. It is based on the definition of the function *PPROD*. Form the table *V∘.×W*, where *V* is the coefficient vector of *N*th order Taylor polynomial of *F* at *A*, as defined above, and *W* is the corresponding vector for *G*. The table for *N* equal to 2 appears in Figure 4.5.4, with the variable *FA* representing the value *F A* and *D̲FA* representing *D̲F A*, and so on.

	GA	*D̲GA*	0.5×*D̲D̲GA*
FA	*FA×GA*	*FA×D̲GA*	0.5×*FA×D̲D̲GA*
D̲FA	*D̲FA×GA*	*D̲FA×D̲GA*	0.5×*D̲FA×D̲D̲GA*
0.5×*D̲D̲FA*	0.5×*D̲D̲FA×GA*	0.5×*D̲D̲FA×D̲GA*	0.25×*D̲D̲FA×D̲D̲GA*

PRODUCT OF TAYLOR POLYNOMIALS
Figure 4.5.4

We see from this table that element 0 of *V PPROD W* is *FA×GA*, which is also the value *H A*. Element 1 of *V PPROD W* is *(FA×D̲GA)+D̲FA×GA* which, according to the Times Rule for Derivatives, is identical to *D̲H A*. Element 2 of *V PPROD W* is (0.5×*D̲D̲FA×GA*)+(*D̲FA×D̲GA*)+0.5×*FA×D̲D̲GA*, which is identical to 0.5×*D̲D̲H A*. Thus 3↑*V PPROD W* is identical to the coefficient vector of the second order Taylor polynomial of

H at *A*. For a larger table, continue this argument through the *N*th element of *V PPROD W*.

The proof of the derivative rule also merits comment: term *I*+1 of a Taylor polynomial of *H* at *A* is

```
(÷!I+1)×(D A)×(ω-A)*I+1
```

where *D* denotes (*I*+1)th derivative of *H*. The derivative of this expression is

```
(÷!I)×(D A)×(ω-A)*I
```

which, since the (*I*+1)th derivative of *H* is the *I*th derivative of *D̲H*, is term *I* of a Taylor polynomial of *D̲H* at *A*. Thus the derivative of the (*N*+1)th order Taylor polynomial of *H* at *A* is also the *N*th order Taylor polynomial of *D̲H* at *A*, which completes the proof of the derivative rule. The proof of the integral rule is similar. ⌸60

<u>Pythagorean Functions</u>. The primary function denoted by 0○ω is defined by the identity

```
0○ω
(1-ω*2)*0.5
```

The graph of 0○ω is the upper half of the circle of radius 1 centered at the origin.

The function 0○ω is called a <u>Pythagorean function</u> because of its relation to the Theorem of Pythagoras; e.g., if the length of the hypotenuse of a right triangle is 1, and if *X* is the length of one leg, then 0○*X* is the length of the other.

The Taylor coefficient function of 0○ω, to be called *T̲C*0, can be defined in terms of the composition rule for Taylor coefficient functions. Namely, the polynomial 1-ω*2 is equivalent to

```
((N+1)↑(1-A*2),(¯2×A),¯1) POLY ω-A
```

and consequently *P*:(α+1)↑(*A*-ω*2),(¯2×ω),¯1 is the Taylor coefficient function of 1-ω*2. Therefore

```
(α T̲SQRT ω) T̲COMP α P ω                    [4.5.5]
```

is the Taylor coefficient function of 0○ω, where *T̲SQRT* is the Taylor coefficient function of *SQRT*:ω*0.5 and is defined as

```
T̲SQRT:(SQRT ω)×(1,×\0.5-ια)÷(ω*ια+1)×!ια+1
```

and *TCOMP* is defined in Ex 4.62b. It is not difficult to show that expression 4.5.5 is equivalent to the following definition of *TC*0:

```
TC0:(α TSQRT ω) TCOMP 0,(¯2×ω),¯1
```

The other primary Pythagorean functions are denoted by 4○ω and ¯4○ω and defined by the identities

```
4○ω                    ¯4○ω                 [4.5.6]
(1+ω*2)*0.5            (¯1+ω*2)*0.5
```

Their graphs and Taylor coefficient functions will be discussed in the exercises.

One reason for introducing the Pythagorean functions in this section is to illustrate a useful application of Taylor coefficient functions, which is the evaluation of high order derivatives. For example, in order to evaluate the fifth derivative of 0○ω for the argument 0.1, we could first produce an expression that defines this derivative, and then evaluate the expression for the specified argument. The first derivative of 0○ω is

```
-ω÷0○ω
```

and the second derivative is

```
-(÷0○ω)+ω÷(0○ω)*3
```

Evidently the expressions that define the higher derivatives become more and more complicated, and it will be quite awkward to produce the expression that defines the fifth derivative of 0○ω. However, consider the vector 5 *TC*0 0.1. Since this vector is the coefficient vector of the fifth order Taylor polynomial of 0○ω at 0.1, it follows from the definition of Taylor polynomials that the element (5 *TC*0 0.1)[5] is the value of the fifth derivative of 0○ω at 0.1, divided by !5. Thus (5 *TC*0 0.1)[5]×!5 is the value of the fifth derivative of 0○ω at 0.1. ⌹64

5 The Definite Integral

5.1 RIEMANN SUMS

The vector `P` is called an increasing partition of the interval `(A≤ω)∧ω≤B` if

```
(A=P[0])∧(B=P[¯1+ρP])∧∧/0<DIFF P
```

For example, the vector `P←1 1.5 1.8 2.3 2.7 3` is an increasing partition of the interval `(1≤ω)∧ω≤3`. Analogously, `P` is called a decreasing partition of the interval `(A≤ω)∧ω≤B` if

```
(B=P[0])∧(A=P[¯1+ρP])∧∧/0>DIFF P
```

For example, the vector `P←3 2.7 2.3 1.8 1.5 1` is a decreasing partition of the interval `(1≤ω)∧ω≤3`.

A partition of the interval `(A≤ω)∧ω≤B` is a vector `P` which is either an increasing or decreasing partition of this interval.

Every vector `P` for which `∧/0=DIFF ×DIFF P` is the partition of some interval, namely the interval with end points `P[0]` and `P[¯1+ρP]`. Consequently no ambiguity will arise if we simply refer to a vector `P` as a partition without specifying the interval it partitions.

The vector `T` is said to populate the partition `P` if `(ρP)=1+ρT` and each element `T[I]` is either between `P[I]` and `P[I+1]` or equal to one of them. Formally, `T` populates `P` if

```
∧/1≠×(T-1↓P)×T-¯1↓P
```

For example, `T←0.5 1 3 4.3` populates `P←0 1 2.5 4 6`, as illustrated in Figure 5.1.1, but `0 0.5 3 4.3` does not.

```
      P←0 1 2.5 4 6
      T←0.5 1 3 4.3
      1≠×(T-1↓P)×T-¯1↓P
1 1 1 1
```
⌹1

T POPULATES P
FIGURE 5.1.1

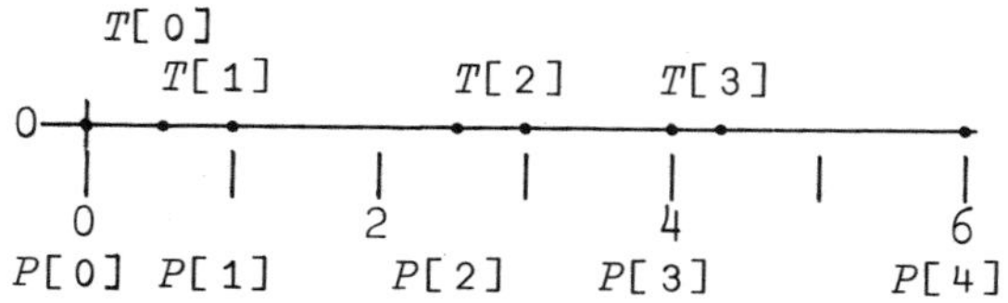

For a function *F*, a partition *P*, and a vector *T* which populates *P*, the scalar defined by the expression

```
+/(F T)×DIFF P
```

is called a Riemann sum of *F* (for the 19th century German mathematician G. F. B. Riemann). Geometrically, if ∧/(0<F T)∧0<DIFF P then this Riemann sum represents a sum of areas of rectangles. For example, consider the function F:ω*2 and the vectors P←¯1 0 2 3 and T←¯1 1.5 3 (Figure 5.1.2). The height of the rectangle labelled 0 is F T[0] and the length of its base is (DIFF P)[0]; the area of this rectangle is (F T[0])×(DIFF P)[0]. In general, the area of the rectangle labelled *I* is (F T[I])×(DIFF P)[I] and the sum of the areas of the three rectangles is +/(F T)×DIFF P. ⌹4

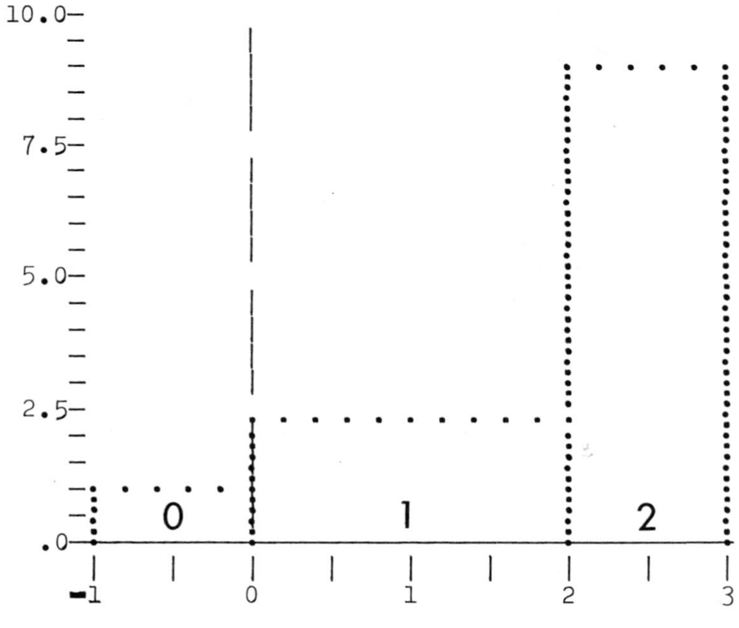

AREA REPRESENTED BY A RIEMANN SUM
FIGURE 5.1.2

There are many interpretations of Riemann sums in applied mathematics, and the one in terms of area is not the most important one. However, it is useful because the concept of area is a familiar one. Other interpretations will be discussed in Section 5.2 and the exercises.

In analyzing Riemann sums it is helpful to examine the vector defined by +\(F T)×DIFF P, all of whose elements are Riemann sums. Continuing the previous example for the function *F*:ω*2:

```
      P
¯1 0 2 3

      T
¯1 1.5 3

      RS←+\(F T)×DIFF P
      RS
1 5.5 14.5
```

RS[0] is the area of the rectangle labelled 0, *RS*[1] is the sum of the areas of the rectangles labelled 0 and 1, and *RS*[2] is the sum of the areas of all three rectangles.

If 0 is appended to the left of the vector +\(*F T*)×*DIFF P*, the result is identical to the vector defined by the expression *SUM* (*F T*)×*DIFF P*. It is now easy to describe the relation between Riemann sums and difference-quotients. Starting with a partition *P* and a vector of values *F T*, form the vector of Riemann sums *RS* as follows:

```
      RS←SUM (F T)×DIFF P
```

Now form the difference-quotient (*DIFF RS*)÷*DIFF P* of the vector of Riemann sums *RS*. The result is the vector of values *F T* with which we started.

```
      (DIFF RS)÷DIFF P
      (DIFF SUM (F T)×DIFF P)÷DIFF P    Definition of RS
      ((F T)×DIFF P)÷DIFF P             Identity 1.1.1
      F T
```

Continuing the previous example:

```
      RS←SUM(F T)×DIFF P
      RS
0 1 5.5 14.5

      (DIFF RS)÷DIFF P                       F T
1 2.25 9                               1 2.25 9
```

Roughly speaking, as a function of its right argument, the difference-quotient function (*DIFF* ω)÷*DIFF* α is the inverse function of the accumulated Riemann sums function *SUM* ω×*DIFF* α. Since difference-quotients are closely related to derivatives and integrals are, again loosely speaking, inverses of derivatives, one should suspect that there is a relationship between integrals and Riemann sums. This relationship is precisely described by the Fundamental Theorems of the Calculus in Section 5.4. ⌹7

5.2 WATER PRESSURE

In this section we will examine the interpretation of Riemann sums in terms of the force exerted by water on a side of a tank. In order to describe this force we will first discuss the concept of pressure at a point.

Assume that the tank is represented in a coordinate system, so that a point in the tank is represented by a vector *P* of its coordinates. Imagine a point *P* in the tank and a cylinder cut vertically through the water in such a way that the point *P* is inside the cylinder. The cylinder encloses an imaginary thin plate which is parallel to the surface of the water and contains the point *P*, as illustrated in Figure 5.2.1.

Suppose that the function *DEPTH* is so defined that for each point *P* in the tank the function value *DEPTH P* is the depth of the point *P* below the surface of the water. If the scalar *AREA* represents the area of the plate described above, then the volume of the part of the cylinder above the plate is identical to *AREA×DEPTH P*. The weight per unit volume of water is called the density of water. If the scalar named *DENSITY* represents the density of water, then the weight of the water in the cylinder above the plate is *DENSITY×AREA×DEPTH P*. Since weight is just another word for the force exerted on the plate by the water, the force on the plate is identical to *DENSITY×AREA×DEPTH P*. The force on the plate per unit area is called the pressure on the plate and is identical to *(DENSITY×AREA×DEPTH P)÷AREA*, that is, *DENSITY×DEPTH P*. Since the pressure on every plate parallel to the surface of the water and containing the point *P* is identical to *DENSITY×DEPTH P*, we define the pressure at the point to be the function *PRES:DENSITY×DEPTH ω*.

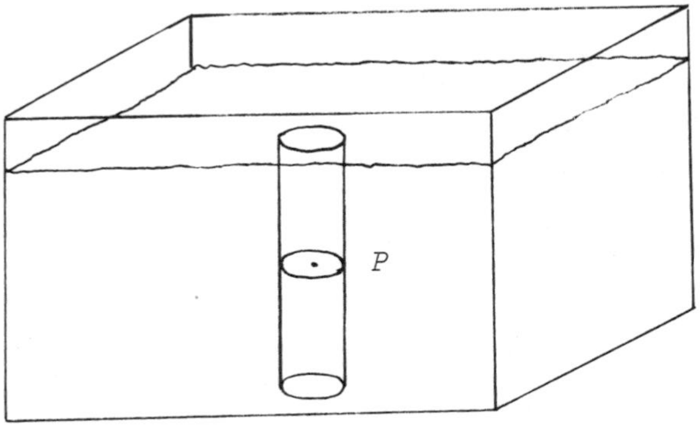

PRESSURE AT A POINT
FIGURE 5.2.1

Thus the force on a plate lying parallel to the surface of the water is *AREA*×*PRES P*, where *AREA* is the area of the plate and *P* is any point on the plate.

We now turn to the problem of determining the force on a side of the tank. Imagine a narrow strip on the side of the tank as illustrated in Figure 5.2.2. Since the strip is not parallel to the surface of the water (the edge of the strip is parallel to the surface of the water), the points on the strip are not all at the same depth and consequently the pressures at the points on the strip vary from point to point. Thus we cannot claim that the force on the strip is identical to the area of the strip times the pressure at a point in the strip. However, since the strip is narrow the depths of any two points in the strip are approximately equal and the pressures at any two points are approximately equal. Thus we can make the reasonable assumption that the force on this narrow strip is approximately equal to the area of the strip times the pressure at any point in the strip.

To approximate the total force on a side of the tank we divide the side into narrow strips, as illustrated in Figure 5.2.3. The scalar *L* represents the length of the side of the tank and *DP* represents the depth of the water in the tank. The vector *D* is such that (0=1↑*D*)∧*DP*=¯1↑*D*, and otherwise the element *D*[*I*] is identical to the depth of the boundary line between the two strips labelled *I*-1 and *I*.

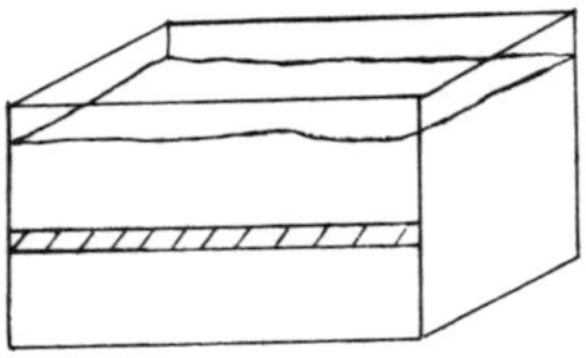

A STRIP ON A SIDE OF THE TANK
FIGURE 5.2.2

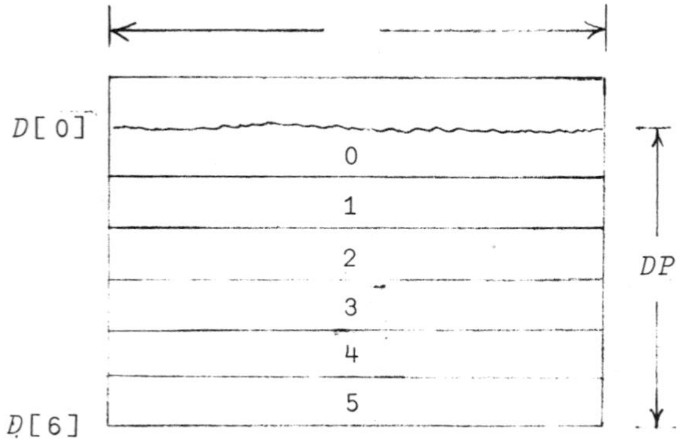

A SUBDIVISION OF A SIDE OF THE TANK
FIGURE 5.2.3

Thus the *I*th element of the vector defined by *L×DIFF D* is identical to the area of the strip labelled *I*. If *Q* is a vector which populates *D*, then for each index *I* the strip labelled *I* contains points at the depth *Q*[*I*]. Thus the *I*th element of the vector defined by *DENSITY×Q×L×DIFF D* is approximately equal to the force exerted by the water on the strip labelled *I*. Finally, the expression

```
+/DENSITY×Q×L×DIFF D
```

approximates the total force on the side of the tank.

If we choose *Q* as *Q*←¯1↓*D*, then the *I*th element of the vector *Q* is the depth of a point on the top edge of the strip labelled *I*, where the pressure at a point is the least for all points in that strip. Evidently then, the value of the above expression for this choice of *Q* is less than the force on the side of the tank. A similar argument shows that the value of the above expression is greater than the force on the side of the tank if *Q* is chosen as *Q*←1↓*D*. If we choose *Q* as *Q*←0.5×(1↓*D*)+¯1↓*D*, then the *I*th element of the vector *Q* is the depth of a point midway between the top edge and the bottom edge of the strip labelled *I*, where the pressure at a point is the average pressure at all points in that strip. It is reasonable to expect that the value of the above expression for this choice of *Q* is a good approximation to the force on the side of the tank. Consider the following sequence for this choice of *Q*:

```
+/DENSITY×L×Q×DIFF D                          [5.2.1]
+/DENSITY×L×0.5×((1↓D)+¯1↓D)×DIFF D
0.5×DENSITY×L×+/(1↓D*2)-¯1↓D*2
0.5×DENSITY×L×-/D[(¯1+ρD),0]*2
0.5×DENSITY×L×DP*2
```

Note that the last expression in this sequence does not contain the vector *D*. This means that for any choice of the vector *D* and for *Q*←0.5×(1↓*D*)+¯1↓*D*, the force on the side of the tank is approximately equal to 0.5×*DENSITY×L×DP**2. One should then expect that the force is actually equal to 0.5×*DENSITY×L×DP**2. This is in fact true, as can be verified by experiments. ⌸13

5.3 ESTIMATING AREAS

Consider a function *F* for which 0≤*F* ω for all arguments in the interval (*A*≤ω)∧ω≤*B*. If *P* is an increasing partition of this interval and if the vector *T* populates *P*, then the value of the Riemann sum

```
+/(F T)×DIFF P
```

is an approximation to the area of the region under the graph of *F* (Figure 5.3.1). The question we are concerned

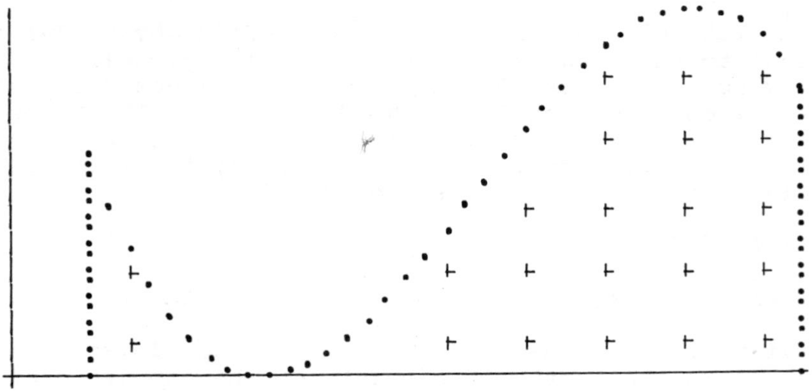

AREA UNDER A GRAPH
FIGURE 5.3.1

with in this section is: how good is the approximation? The answer is easy to formulate if *F* is also a monotonic function. If the scalar *AREA* represents the area of the region and if *F* is monotonic on the interval (*A*≤ω)∧ω≤*B*, then

```
(|AREA -+/(F T)×DIFF P)≤|(-/F B,A)×⌈/|DIFF P          [5.3.2]
```

For example, consider the problem to determine to 1 decimal place accuracy the area of the region bordered by the graph of the function *F*:ω⋆2, the horizontal axis, and the vertical lines at 1 and 3 (Figure 5.3.3). In view of

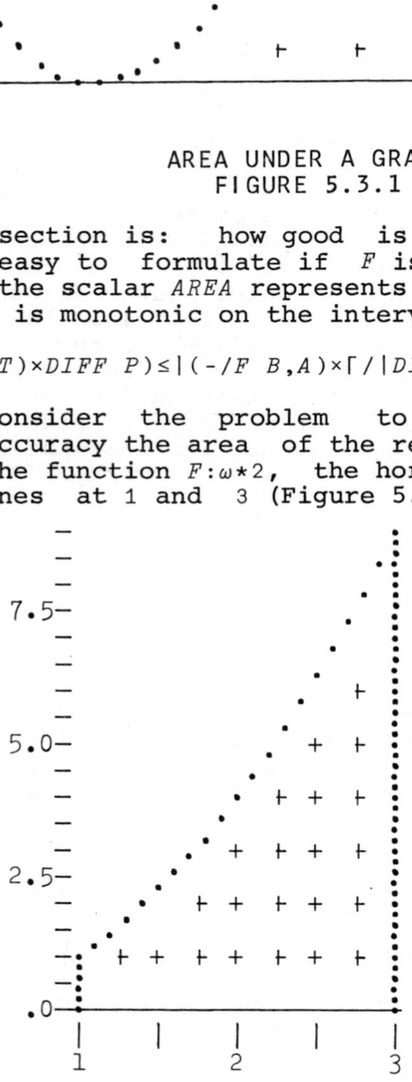

AREA UNDER THE GRAPH OF ω⋆2
FIGURE 5.3.3

Inequality 5.3.2, the problem reduces to determining an increasing partition *P* of the interval `(1≤ω)∧ω≤3` for which `0.05>(-/F 3 1)×⌈/|DIFF P`. Since `8=-/F 3 1`, we may choose a partition *P* for which

```
      (⌈/|DIFF P)<0.05÷8
```

For example, the partition

```
      P←1+(⍳326)×2÷325
```

will do.

```
      P[0,¯1+⍴P]
1 3
      0.05>(-/F 3 1)×⌈/|DIFF P
1
```

Now choose any vector *T* which populates *P*, say:

```
      T←0.5×(1↓P)+¯1↓P
```

Then

```
      +/(F T)×DIFF P
8.66666035
```

is the required approximation. ⊞19

Inequality 5.3.2 is the result of four other inequalities, all of which are obvious consequences of the fact that *F* is monotonic. We will present these inequalities for non-decreasing functions. The reader should determine the analogous inequalities for non-increasing functions.

First of all, if *F* is non-decreasing on the interval `(A≤ω)∧ω≤B`, then

```
      AREA≤+/(F 1↓P)×DIFF P                    [5.3.4]

      AREA≥+/(F ¯1↓P)×DIFF P                   [5.3.5]
```

The validity of these inequalities can be seen from the appropriate graphs in Figures 5.3.6 and 5.3.7.

Next, since *F* is a non-decreasing function the relation `∧/T≥¯1↓P` implies the relation `∧/(F T)≥F ¯1↓P` and equivalently `∧/(F T)≥¯1↓F P`. Therefore

```
      (+/(F T)×DIFF P)≥+/(¯1↓F P)×DIFF P
```

and therefore

```
      (-+/(F T)×DIFF P)≤-+/(¯1↓F P)×DIFF P     [5.3.8]
```

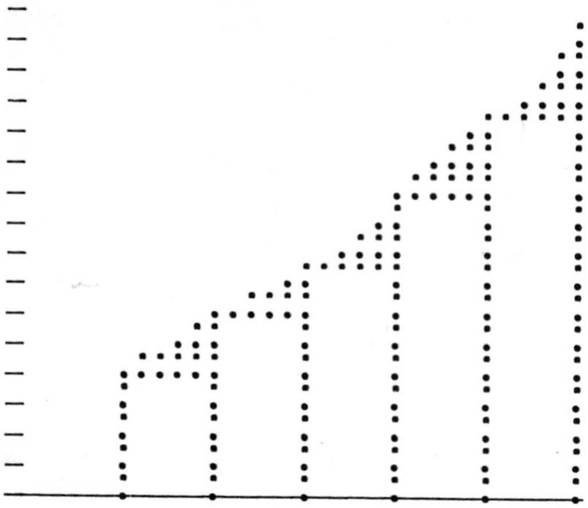

COMPARISON OF *AREA* TO
+/(*F* ¯1↓*P*)×*DIFF P*
FIGURE 5.3.6

Similarly:

(-+/(*F T*)×*DIFF P*)≥-+/(1↓*F P*)×*DIFF P* [5.3.9]

Summing the expressions in 5.3.4 and 5.3.8 yields the inequality

(*AREA*-+/(*F T*)×*DIFF P*)≤+/(*DIFF F P*)×*DIFF P*

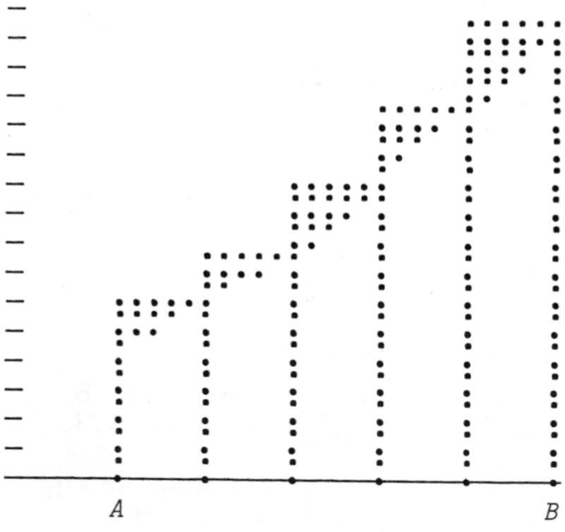

COMPARISON OF *AREA* TO
+/(*F* 1↓*P*)×*DIFF P*
FIGURE 5.3.7

Summing the expressions in 5.3.5 and 5.3.9 yields the inequality

```
(AREA-+/(F T)×DIFF P)≥-+/(DIFF F P)×DIFF P
```

The last two inequalities can be combined into one:

```
(|AREA-+/(F T)×DIFF P)≤|+/(DIFF F P)×DIFF P
```

Therefore:

```
(|AREA-+/(F T)×DIFF P)≤(|+/DIFF F P)×⌈/|DIFF P      [5.3.10]
```

Last of all, we have the sequence:

```
+/DIFF F P
+/(1↓F P)-¯1↓F P                          Def'n of DIFF
-/F P[(¯1+ρP),0]
-/F B,A                                   Def'n of P
```

Substituting the expression `-/F B,A` for the expression `+/DIFF F P` in 5.3.10 yields 5.3.2.

<u>Signed</u> <u>Area</u>. As we have seen, the Riemann sum

```
+/(F T)×DIFF P
```

can be interpreted in terms of area if *P* is an increasing partition and if `0≤F T`. More generally, if the values of *F* are not all non-negative, the Riemann sum can be interpreted in terms of signed area.

For example, consider the function `F:1-ω*2` and the vectors `P←.2×ι11` and `T←.1+.2×ι10` (Figure 5.3.11). For an index *I* such that `(0≤I)∧I≤4`, the *I*th element of the vector defined by the expression `(1-T*2)×DIFF P` is identical to the area of the rectangle labelled *I*. However, if `(5≤I)∧I≤9` then the *I*th element of this vector is identical to ¯1 times the area of the rectangle labelled *I*. Thus the value of the Riemann sum defined by `+/(F T)×DIFF P` is identical to the sum of areas of the rectangles labelled 0 through 4 minus the sum of the areas of the rectangles labelled 5 through 9. This value is called the signed area of the region composed of the rectangles. ⊞35

In general, the <u>signed</u> <u>area</u> of a region in a coordinate plane is the area of that part of the region that lies above the horizontal axis minus the area of that part of the region that lies below the horizontal axis. As an illustration, consider the graph of a function *F* in Figure 5.3.12. The region of the coordinate plane bordered by the graph of *F*, the horizontal axis, and the vertical lines at *A* and *B* is composed of the three regions labelled *R*1,*R*2,*R*3. The area of the region is identical to the sum of the areas of *R*1,*R*2 and *R*3. The signed area of the region is the sum of the areas of *R*1 and *R*3 minus the area of *R*2.

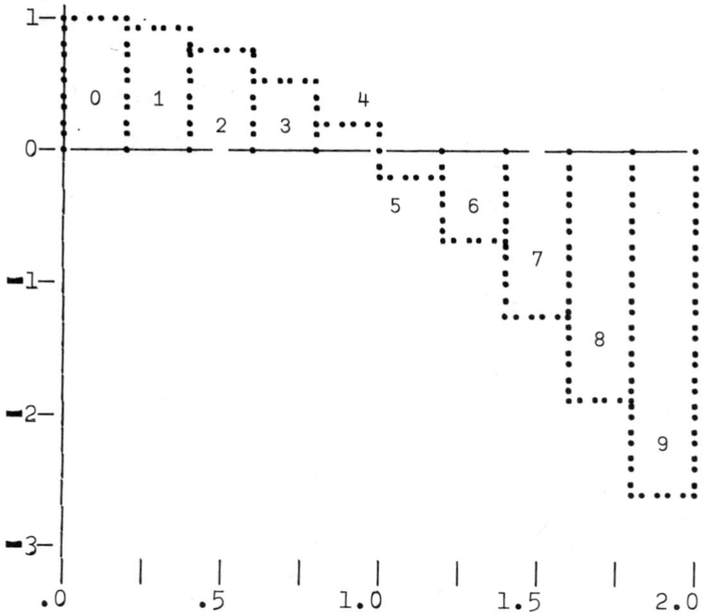

SIGNED AREA REPRESENTED BY A RIEMANN SUM
FIGURE 5.3.11

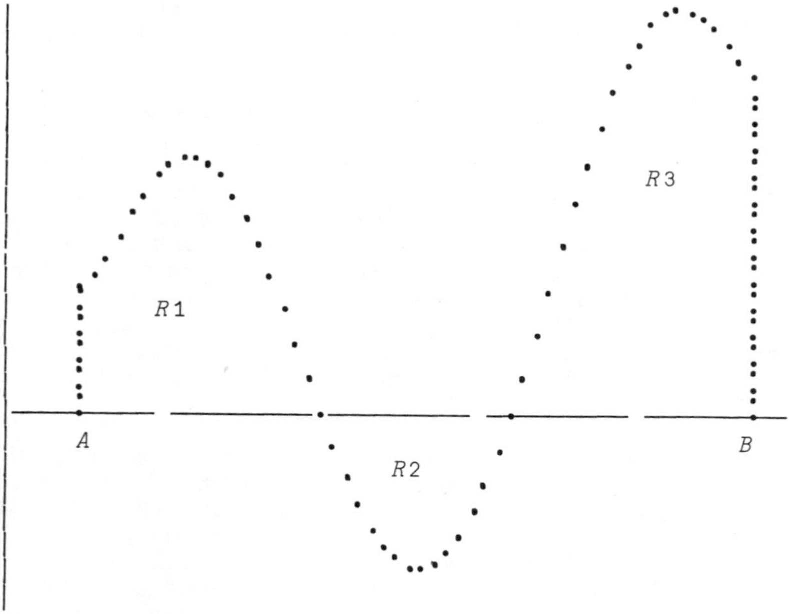

SIGNED AREA UNDER A GRAPH
FIGURE 5.3.12

Since Riemann sums approximate signed areas of regions in a coordinate plane, the question again arises as to how good is a particular approximation. A careful examination of the proof of 5.3.2 will show that this inequality remains valid for signed areas and monotonic functions. ⊞37

Note that if *P* is a decreasing partition of the interval (*A*≤ω)∧ω≤*B*, then a Riemann sum of the form +/(*F T*)×*DIFF P* is an approximation to minus the signed area of the region in a coordinate plane bordered by the graph of *F*, the horizontal axis, and the vertical lines at *A* and *B*. To see that this is true consider the sequence of identities:

```
+/(F T)×DIFF P
+/⌽(F T)×DIFF P
+/(⌽ F T)×⌽DIFF P
+/(F ⌽T)×-DIFF ⌽P
-+/(F ⌽T)×DIFF ⌽P
```

Evidently the partition defined by ⌽*P* is an increasing partition of the interval (*A*≤ω)∧ω≤*B* and consequently the Riemann sum +/(*F* ⌽*T*)×*DIFF* ⌽*P* is an approximation to the signed area of the region described above. Therefore, according to the above sequence, the Riemann sum +/(*FM T*)×*DIFF P* is an approximation to minus the signed area of this region.

5.4 DEFINITE INTEGRALS

Let *F* be a monotonic function on the interval (*A*≤ω)∧ω≤*B* and define the function

```
MD:2×(|-/F B,A)×⌈/|(DIFF α),DIFF ω
```

Suppose that *P* and *Q* are increasing partitions of this interval and that *T* and *R* are vectors which populate *P* and *Q* respectively. By an argument similar to the one in Section 5.3 used to prove 5.3.2 we can prove:

```
(P MD Q)≥|(+/(F T)×DIFF P)-+/(F R)×DIFF Q
```

According to this inequality there is a constant *V* approached by the Riemann sums +/(*F T*)×*DIFF P* when the partitions *P* are chosen so that the values of ⌈/|*DIFF P* are close to zero. To see this we need only show how to determine any specified number of digits in the decimal representation of *V*. For example, according to the above inequality the values of any two Riemann sums +/(*F T*)×*DIFF P* and +/(*F R*)×*DIFF Q* agree to 10 decimal digits for increasing partitions *P* and *Q* for which 0.5*E*¯10>*P MD Q*. Evidently the first ten decimal digits of *V* must be the same as those of any one of these Riemann sums. The value *V* is clearly a function of the endpoints *A* and *B* and is called the value of the *definite integral* of *F* from *A* to *B*. Evidently *V* is the signed area of the region in a plane bordered by the graph of *F*, the horizontal axis, and the vertical lines at *A* and *B*. Note, however, that the definition of *V* is independent of this particular geometric interpretation, just as the definition of derivatives in terms of cline functions is independent of the geometric interpretation in terms of slopes of tangent lines and of the physical interpretation in terms of velocity.

The Definition of the Definite Integrals. To each function of bounded variation *F* (that is, *F* is the sum of two monotonic functions) there is an associated dyadic function named $\underline{S}F$ and called the *definite integral of F*, which is implicitly defined as follows. Each function value *A* $\underline{S}F$ *A* is defined to be zero. If *A≠B*, then for partitions *P* such that (*A*=1↑*P*)∧*B*=¯1↑*P* and vectors *TP* which populate *P* the function value *A* $\underline{S}F$ *B* is defined to be the value approached by the Riemann sums +/(*F TP*)×*DIFF P* when ⌈/|*DIFF P* is close to zero.

According to the above definition, if *A*<*B* then *A* $\underline{S}F$ *B* is identical to the signed area of the region in a plane bordered by the graph of *F*, the horizontal axis, and the vertical lines at *A* and *B*. In view of the concluding paragraph of Section 5.3, if *A*>*B* the function value *A* $\underline{S}F$ *B* is minus this signed area. Therefore:

A $\underline{S}F$ *B* [5.4.1]
-*B* $\underline{S}F$ *A*

The interpretation of definite integrals in terms of signed area suggests another important identity. Evidently the area of the region *R*1 illustrated in Figure 5.4.2 is identical to the sum of the areas of *R*2 and *R*3. In terms of the definite integral $\underline{S}F$ of *F*, this relation between signed areas is expressed by the identity:

A $\underline{S}F$ *C* [5.4.3]
(*A* $\underline{S}F$ *B*)+*B* $\underline{S}F$ *C*

The following rules for definite integrals are more easily demonstrated using the definition of definite integrals by Riemann sums.

The Plus Rule for Definite Integrals. For functions *F* and *G* and their definite integrals $\underline{S}F$ and $\underline{S}G$ respectively, if *H* ω is equivalent to (*F* ω)+*G* ω then

α $\underline{S}H$ ω
(α $\underline{S}F$ ω)+α $\underline{S}G$ ω

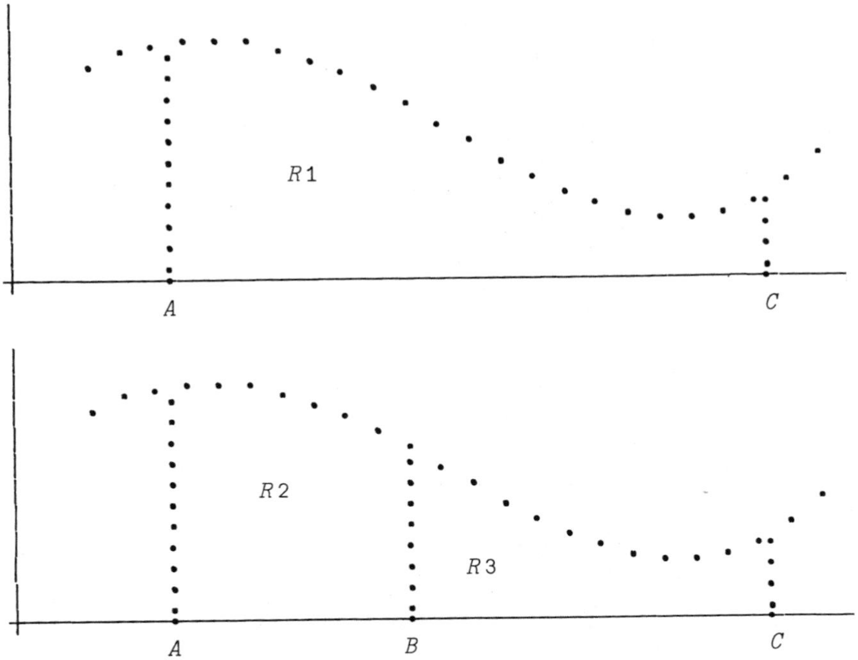

GEOMETRIC INTERPRETATION OF IDENTITY 5.4.3
FIGURE 5.4.2

To demonstrate this rule consider the following sequence of identities:

```
+/(P T)×DIFF P
+/((F T)+G T)×DIFF P
+/((F T)×DIFF P)+(G T)×DIFF P
(+/(F T)×DIFF P)++/(G T)×DIFF P
```

For partitions *P* such that (*A*=1↑*P*)∧*B*=¯1↑*P* and vectors *T* populating *P*, the values of the first expression in this sequence approach *A S̲H B* as the values ⌈/|*DIFF P* approach 0. At the same time the values of the last expression in this sequence approach (*A S̲F B*)+*A S̲G B*. Since the expressions in the above sequence are equal it follows that *A S̲P B* is equal to (*A S̲F B*)+*A S̲G B*.

<u>The Scale Rule for Definite Integrals</u>. For a function *F* with its definite integral *S̲F* and a scale function *SC*:*C*+*D*×ω for which 0≠*D*, if *H* ω is equivalent to *F SC* ω then

```
α S̲H ω
(÷D)×(SC α) S̲F SC ω.
```

To demonstrate this rule consider the sequence:

```
+/(H T)×DIFF P
+/(F SC T)×DIFF P
(÷D)×+/(F SC T)×DIFF D×P
(÷D)×+/(F SC T)×DIFF C+D×P
(÷D)×+/(F SC T)×DIFF SC P
```

For partitions *P* such that (A=1↑P)∧B=¯1↑P and vectors *T* populating *P*, the values of the first expression in this sequence approach A S̲H B as the values ⌈/|DIFF P approach zero. At the same time the vector defined by SC P is a partition of the interval with end points SC A and SC B, the vector SC T populates SC P, and therefore the values of the last expression approach (÷D)×(SC A) S̲F SC B. Since the values of the expressions in this sequence are equal it follows that A S̲H B is equal to (÷D)×(SC A) S̲F SC B. ⊞47

Fundamental Theorems of the Calculus. The fundamental theorems of the calculus describe the relation between definite integrals and integrals (that is, anti-derivatives). To state and prove these theorems we will use the concept of *continuous functions*. Roughly speaking, the function *F* is said to be *continuous at the argument* *A* if there is no break in the graph of *F* at *A* and *discontinuous at* *A* if there is a break in the graph. In other words, *F* is continuous at *A* if the values F X approach the value F A as *X* is chosen closer and closer to *A*. For example, the function ×ω is discontinuous at zero and ω*2 is continuous at zero. In fact, every polynomial is continuous everywhere.

The First Fundamental Theorem. Suppose that *F* is a monadic scalar function with the definite integral S̲F and define G:C S̲F ω for a scalar *C*. Then the value D̲G X of the derivative of *G* is equal to F X for each argument *X* at which *F* is continuous. In particular, if *F* is continuous at every argument then G ω is an integral of F ω.

The Second Fundamental Theorem. If *F* is a monadic scalar function with the definite integral S̲F, if *IF* is an integral of *F*, and if *F* is continuous at every argument in an interval (A≤ω)∧ω≤B, then:

```
C S̲F D
-/IF D,C
```

for every pair of arguments *C* and *D* in this interval.

The Second Fundamental Theorem provides a way of producing explicit definitions of definite integrals in terms of explicit definitions of integrals. For example, consider the polynomial F:1 ¯3 6 POLY ω. The function IF:0 1 ¯1.5 2 POLY ω is an integral of *F*. Consequently, according to the Second Fundamental Theorem, the definite integral *S̲F* can be defined by *S̲F*:-/*IF* ω,α. In particular, the signed area of the region bordered by the graph of *F*, the horizontal axis, and the vertical lines at 1 and 4 is -/*IP* 4 1, or 108-1.5, or 106.5.

More generally, the definite integral of the polynomial *F*:*C POLY* ω is

```
S̲F:-/(0 INTEG C) POLY ω,α
```

Note that any constant can be used in place of 0 in the definition of *S̲F*.

It is sometimes useful for the coefficient vector to be an explicit argument of a function producing values of definite integrals of polynomials. This is accomplished by the function

```
DEFINT:-/(0 INTEG ω) POLY ⌽α
```

For example, the value *A S̲F B* of the definite integral function defined above is equal to (*A*,*B*) *DEFINT C*. ⊟50

Analogously, the First Fundamental Theorem provides a way of producing explicit definitions of integrals in terms of explicit definitions of definite integrals. It is very unusual to know an explicit definition of a definite integral function without knowing explicit definitions of integrals, but approximate explicit definitions of definite integrals can be defined in terms of Riemann sums of fitting polynomials. The First Fundamental Theorem can then be used to produce approximate explicit definitions of integrals. The details of the approximation procedures are discussed in the next section.

The proof of the First Fundamental Theorem for a monotonic function *F* is based on Identity 5.4.3 and the definition of derivatives in terms of cline functions. Substituting *C* for *A*, *X* for *B*, and *X*+*S* for *C* in 5.4.3 yields the following addition formula for *G*:*C S̲F* ω:

```
G X+S
(G X)+X S̲F X+S
(G X)+S×(X S̲F X+S)÷S
```

Consequently

```
G ω+α
(G ω)+α×(ω S̲F ω+α)÷α
```

and (ω *S̲F* ω+α)÷α is equivalent to the cline function α *C̲G* ω when α≠0.

Since F is monotonic, the function value $X\ \underline{S}F\ X+S$ lies between $S\times F\ X$ and $S\times F\ X+S$, as illustrated in Figure 5.4.4. Thus the value of the cline function $S\ \underline{C}G\ X$ lies between $F\ X$ and $F\ X+S$.

If F is continuous at X then the values $F\ X+S$ approach the value $F\ X$ as S approaches zero. Thus since the values $S\ \underline{C}G\ X$ are between $F\ X$ and $F\ X+S$, these values must also approach $F\ X$ as S approaches zero. Therefore the value $0\ \underline{C}G\ X$ is defined and is equal to $F\ X$, which was to be proved.

The Second Fundamental Theorem follows from the First Fundamental Theorem. According to the latter, the function $G{:}C\ \underline{S}F\ \omega$ is an integral of F. Since IF is also an integral of F there is a constant S for which:

$$G\ \omega$$
$$S+IF\ \omega$$

In particular, $G\ C$, which is zero, is equal to $S+IF\ C$, and hence S is equal to $-H\ C$. Therefore $G\ \omega$ equals $(IF\ \omega)-IF\ C$ and consequently

$$C\ \underline{S}F\ D$$
$$G\ D$$
$$(IF\ D)-IF\ C$$
$$-/IF\ D,C$$

We have proved the fundamental theorems for monotonic functions F. It follows from the sum rules for derivatives and definite integrals that the theorems are also true of

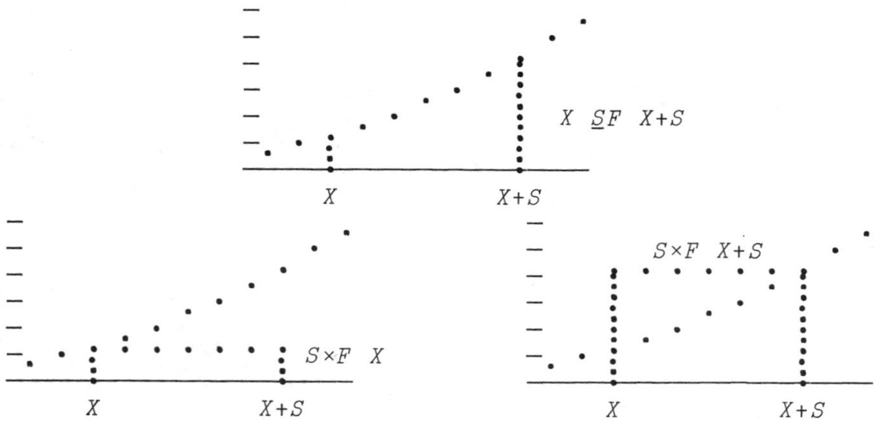

COMPARISON OF AREAS
FIGURE 5.4.4

functions that are the sum of two monotonic functions, and therefore true for functions of bounded variation.

<u>The Definite Integral of the Function</u> *ω⋆N*. If *N* is a non-negative integer then the First Fundamental Theorem can be proved for the function *P*:*ω⋆N* directly from Riemann sums. For a scalar *D* for which 0≠*D* and a positive integer *K* the vector defined by the expression

```
(⍳K+1)×S←D÷K
```

is a partition of the interval with end points 0 and *D*. Also, the vector *S*×⍳*K* populates the partition *S*×⍳*K*+1. Therefore, since ∧/*S*=*DIFF S*×⍳*K*+1, the expression

```
S×+/P S×⍳K                                    [5.4.5]
```

is a Riemann sum of *P* on the interval with end points 0 and *B*. Moreover, as the integer *K* is assigned large values, the values of the scalar defined by ⌈/|*DIFF* (⍳*K*+1)×*S* are small. Therefore, according to the definition of the definite integral S̲*P* of *P*, the values of the Riemann sums defined by 5.4.5 approximate the function value 0 S̲*P D* for large values of *K*. We can use the Riemann sums defined by 5.4.5 to prove that:

```
0 S̲P D                                        [5.4.6]
(D⋆N+1)÷N+1
```

Since the function (*ω⋆N*+1)÷*N*+1 is an integral of *P*, 5.4.6 proves that *G*:0 S̲*P ω* is an integral of *P*.

For example, if *N* is 1 we have the following sequence of identities for the Riemann sum defined by 5.4.5:

```
S×+/P S×⍳K                                    [5.4.7]
S×+/(S×⍳K)⋆1
S×+/S×⍳K
(S⋆2)×+/⍳K
((D⋆2)÷K⋆2)×+/⍳K               Definition of S
(D⋆2)×(+/⍳K)÷K⋆2
```

For the scalar defined by +/⍳*K* we have the familiar identity

```
+/⍳K
0.5×K×¯1+K
```

from which the following sequence of identities follows:

```
+/⍳K                                          [5.4.8]
0.5×K×¯1+K
(¯0.5×K)+0.5×K⋆2
+/0 ¯0.5 0.5×K⋆0 1 2
```

Combining these sequences yields:

```
      S×+/P S×⍳K
      (D*2)×(+/⍳K)÷K*2                      Sequence 5.4.7
      (D*2)×(+/0 ¯0.5 0.5×K*0 1 2)÷K*2      Sequence 5.4.8
      (D*2)×+/0 ¯0.5 0.5×K*¯2 ¯1 0
```

As `K` is assigned large values the vector defined by the expression `K*¯2 ¯1 0` assumes values close to `0 0 1`. In turn, the scalar defined by `+/0 ¯0.5 0.5×K*¯2 ¯1 0` assumes values close to `+/0 ¯0.5 0.5×0 0 1`, or equivalently `0.5`. Finally, the scalar defined by `(D*2)×+/0 ¯0.5 0.5×K*¯2 ¯1 0` assumes values close to `0.5×D*2`. Therefore, if `N` is `1`:

```
      0 S̲P D
      0.5×D*2
```

which agrees with 5.4.6.

5.5 APPROXIMATIONS OF DEFINITE INTEGRALS

If an explicit definition of a definite integral function is not known, we can appeal to the definition of definite integrals in terms of Riemann sums to produce approximate values of the function. For example, consider the function of `LD:÷ω` and its definite integral function `S̲LD`. An approximation to the function value `1 S̲LD 3` can be produced as follows:

```
      ⎕←P←1+0.2×⍳11
1 1.2 1.4 1.6 1.8 2 2.2 2.4 2.6 2.8 3

      6⍕+/(LD 0.5×(1↓P)+¯1↓P)×DIFF P
1.097142
```

We will be able to check in Chapter 7 that `1.098612` is an approximation to `1 S̲LD 3` which is accurate to six decimal places.

The approximate value produced in the above example could be improved by using a partition `P` for which the value of the scalar defined by `⌈/|DIFF P` is close to zero. However, the approximation can also be improved by using the same partition with an approximation procedure other than Riemann sums.

For a monadic scalar function `F` and its definite integral function `S̲F`, approximations to the function values `A S̲F B` can be produced using fitting polynomials. Choose a partition `P` of the interval with end points `A` and `B` for which `A=P[0]` and define the vector `C` as follows:

```
      C←P CFP F P
```

The vector *C* is the coefficient vector of the fitting polynomial for the vector of arguments *P* and the vector of values *F P*. In particular:

```
∧/(F P)=C POLY P
```

If this fitting polynomial is a good approximation to the function *F* on the interval with end points *A* and *B*, it is reasonable to expect that the scalar defined by the expression (*A*,*B*) *DEFINT C* is a good approximation to the function value *A S̲F B*. That is,

```
P[0,¯1+ρP] DEFINT (F P)⌹(ρP) VAN P                    [5.5.1]
```

is an approximation to the function value *A S̲F B*. Using this expression as a guide, we define the approximate definite integral function

```
AS̲:α[0,¯1+ρα] DEFINT α CFP ω
```

Continuing the previous example:

```
      6⍕P AS̲ LD P
1.098613
```

The function *AS̲* is a linear function of its right argument, and therefore can be expressed as a +.× inner product as follows:

```
E̲AS̲:(α AS̲ ω[;0]),α E̲AS̲ 0 1↓ω : 0=¯1↑ρω : ι0

COEFF:ω E̲AS̲ ID ρω

α AS̲ ω                                                 [5.5.2]
(COEFF α)+.×ω
```

Continuing the previous examples:

```
      6⍕(COEFF P)+.×LD P
1.098613
```

For scalars *C* and *D* and partitions *P* we have the identity:

```
COEFF C+D×P                                            [5.5.3]
D×COEFF P
```

which is a reflection of the Scale Rule for Definite Integrals. For example:

```
      COEFF 3.5+0.1×1 2 3 4
0.0375 0.1125 0.1125 0.0375

      0.1×COEFF 1 2 3 4
0.0375 0.1125 0.1125 0.0375
```

The importance of Identity 5.5.3 will be seen below.

If the partition *P* contains more than a few elements then in practice it can be imprudent to use *P* as the vector of arguments for a fitting polynomial, and therefore a variation of the approximation method described above must be devised. The commonly used variation is to produce approximations as above for sets of arguments consisting of small numbers of consecutive elements of the partition *P*, and then, according to Identity 5.4.3, sum the results. The preceding example will illustrate this procedure.

Consider the function

```
SPI:(⍳α)∘.+(α-1)×⍳÷/¯1+ω,α
```

(for subpartition indices). For example:

```
      3 SPI 9
0 2 4 6
1 3 5 7
2 4 6 8
```

The matrix `3 SPI ⍴P` is used to arrange the elements of the partition *P* in a matrix.

```
      Q←P[3 SPI ⍴P]
      1⍕Q
1.0 1.4 1.8 2.2 2.6
1.2 1.6 2.0 2.4 2.8
1.4 1.8 2.2 2.6 3.0
```

Each column of the matrix *Q* is a partition, called a <u>subpartition</u> of *P*, and consequently for each index *I*

```
      (COEFF Q[;I])+.×LD Q[;I]
```

is approximately equal to `Q[0;I] SLD Q[2;I]`. According to Identity 5.5.3, each vector `COEFF Q[;I]` is identical to `0.2×COEFF ⍳3`. Consequently the above scalar is equal to `0.2×(COEFF ⍳3)+.×LD Q[;I]`, which is element *I* of the vector

```
      0.2×(COEFF ⍳3)+.×LD Q
```

Finally, since the right end of `Q[;I]` coincides with the left end of `Q[;I+1]`, it follows from Identity 5.4.3 that `+/0.2×(COEFF ⍳3)+.×LD Q` is approximately equal to `1 SLD 3`. Thus:

```
      6⍕0.2×+/(COEFF ⍳3)+.×LD Q
1.098661
```

In general, if *P* is a <u>uniform</u> partition of the interval with end points *A* and *B* (that is, there is a scalar *S* for which `∧/S=DIFF P`), and if *N* is an integer for which

`÷/¯1+(ρP),N` is also an integer, then the scalar defined by the expression

```
S×+/(COEFF ιN)+.×F P[N SPI ρP]                    [5.5.4]
```

is an approximation to the function value `A SF B`.

According to 5.5.4, only the function values `COEFF ιN` are needed for these approximations methods. Table 5.5.5 is a list of several of these function values.

N	COEFF ιN
2	1 1÷2
3	1 4 1÷3
4	3 9 9 3÷8
5	14 64 24 64 14÷45

COEFFICIENTS FOR APPROXIMATE INTEGRATION
TABLE 5.5.5

The approximation method that uses `COEFF ι2` is called the trapezoid method, while the method that uses `COEFF ι3` is known as Simpson's method. In general, the method using `COEFF ιN+1` is said to be of order *N*. The *N*th order method is applied to a function *F* by the function

```
SAF:(-/ω[1 0])×+/CF+.×F ω[(ρCF) SPI ρω]
```

where `CF←COEFF N+1`. For example:

```
      SALD:(-/ω[1 0])×+/CF+.×LD ω[(ρCF) SPI ρω]
      CF←COEFF ι3
      P
1 1.2 1.4 1.6 1.8 2 2.2 2.4 2.6 2.8 3
      6⍕SALD P
1.098661
```

Note that the coefficients `CF←COEFF N+1` are computed once and for all, and are then used in all subsequent applications of the *N*th order method.

When applying the *N*th order method it is convenient to use a function that produces uniform partitions *P* of specified intervals for which `¯1+ρP` is a multiple of `¯1+ρCF`. For that purpose we define

```
PART:α[0]+((-/⌽α)÷N8-1)×ιN8←1+ω×¯1+ρCF
```

For example:

```
      CF←COEFF ι3
      1 3 PART 5
1 1.2 1.4 1.6 1.8 2 2.2 2.4 2.6 2.8 3
```

Note that according to the definition of *PART*, the length of its values (*A*,*B*) *PART K* increase, and therefore ⌈/|*DIFF* (*A*,*B*) *PART K* decrease, as the value of *K* increases. Using the function *PART*, we define

```
S̲BF:S̲AF α PART ω
```

so that for each integer *K*, (*A*,*B*) *S̲BF K* is approximately equal to *A S̲F B*; evidently, the approximations should improve for larger and larger values of *K*.

As always, the question arises as to the accuracy of the approximations. In practice, we apply one of the approximation methods for a sequence of partitions *P*0, *P*1, *P*2, etc., where ⌈/|*DIFF P*0 is greater than ⌈/|*DIFF P*1, which is in turn greater than ⌈/|*DIFF P*2, and so on. If we want, say, six decimal place accuracy then as with Newton's method we stop when two successive approximations agree to six decimal places.

Using this stopping rule as a guide, the definite integral function *S̲F* can be defined for practical purposes as follows:

```
S̲STOPF:0.5E¯6>|(ω S̲BF 2×α)-ω S̲BF α
S̲CF:ω S̲CF 2×α : ω S̲STOPF α : ω S̲BF 2×α
S̲F:(α,ω) S̲CF 5
```

(the right argument 5 of *S̲CF* in the definition of *S̲F* provides a partition for the initial approximation, and is chosen arbitrarily). For example, defining the above functions for the function *LD* and setting

```
C̲F̲←COEFF ⍳7
```

we have:

```
      6⍕1 S̲LD 3
1.098613
```

Since *S̲F* is approximately the definite integral function of *F*, it follows from the First Fundamental Theorem that the functions *A S̲F ω* for constants *A* are approximate integrals of *F ω*. For example, the functions *A S̲LD ω* are approximately integrals of *LD ω*. Up to now we have been unable to explicitly define integrals of *LD*, because it is the one function of the form *ω*⋆*N* whose integrals are not of the form *C*+(*ω*⋆*N*+1)÷*N*+1 for constants *C*. ⌹76

6

Elementary Transcendental Functions

6.1 INTRODUCTION

We will now construct the elementary transcendental functions, which consist of growth and decay functions, circular functions, and hyperbolic functions. Growth and decay functions are proportional to their derivatives. The circular functions are commonly called the sine and cosine functions, while the hyperbolic functions are called the hyperbolic sine and hyperbolic cosine functions. The circular and hyperbolic functions are proportional to their second derivatives. They are called transcendental functions because they are not polynomials or fractional powers of polynomials. They can, however, be constructed from approximating polynomials, and we begin with a theorem which is used to estimate the errors in these approximations.

Quotients and Products. For the function *QT*:(1↓ω)÷¯1↓ω and a vector *V* the elements of the vector *QT V* are quotients of successive elements of *V*. For example:

```
      QT 1 2 6 18 72
2 3 3 4
```

The function *PRD*:1,×\α is called the augmented times scan. For example:

```
      V                                  PRD V
¯1 2 3 5 2 ¯2                      1 ¯1 ¯2 ¯6 ¯30 ¯60 120
```

For vectors *V* such that ∧/0≠*V*, scalars *C*, and the functions *QT* and *PRD* we have the identities:

```
      V                            V                          [6.1.1]
      V[0]×PRD QT V                QT C×PRD V
```

For example:

```
      V                                  C
¯1 2 3 5 2 ¯2                      4

      V[0]×PRD QT V                      QT C×PRD V
¯1 2 3 5 2 ¯2                      ¯1 2 3 5 2 ¯2
```

The principal application of the functions *QT* and *PRD* in this text is the analysis of the terms of polynomials. ⌹1

<u>Theorem</u>. For a vector *C*, a scalar *B* such that `B>0` and a non-negative integer *K* define

```
M←B×⌈/QT K↓|C
```

Then if `1>M`,

```
(|((-ρC)↑K↓C) POLY X)≤(|C[K]×B*K)÷1-M                [6.1.2]
```

for all arguments *X* for which `B≥|X`.

The proof of this theorem will be presented after the following example.

As an illustration of the application of theorem 6.1.2, consider the problem of producing a table of values of the polynomial with coefficient vector `÷!ι100` for arguments *X* such that `2≥|X`. Evidently it is a long and tedious task to produce even a few values of this polynomial. The problem is much more manageable if only limited accuracy is required of the table, say to three decimal places. To produce such a table we could first determine an integer *L* for which

```
0.0005>|((÷!ι100) POLY X)-(÷!ιL) POLY X
```

whenever `2≥|X`. If *L* is comparatively small then it will not be difficult to produce a table of values for the polynomial with coefficient vector `÷!ιL`. This table will serve as a table of values for the polynomial with coefficient vector `÷!ι100` within the required accuracy. Since

```
((÷!ι100) POLY X)-(÷!ιL) POLY X
(¯100↑L↓÷!ι100) POLY X
```

the problem comes down to producing an integer *L* for which

```
0.0005>|(¯100↑L↓÷!ι100) POLY X
```

whenever `2≥|X`.

The integer *L* described in the preceding paragraph can be produced from the statement of Theorem 6.1.2. We first substitute `÷!ι100` for *C*. We will substitute the value `2` for *B* later. Next we determine the scalar *M* defined in the statement of the theorem. According to the definition of the successive quotient function *QT*, element *I* of the vector `QT ÷!ι100` is identical to `÷I+1`. Therefore:

```
B×⌈/|K↓QT÷!ι100                                      [6.1.3]
B×⌈/|K↓÷1+ι99
B÷K+1
```

Therefore *M* is *B*÷*K*+1. In particular, 1>*M* if *K*≥*B*. Moreover, we have the following identity for the expression on the right hand side of the inequality in the statement of Theorem 6.1.2:

```
(|C[K]×B*K)÷1-M                                  [6.1.4]
(B*K)÷(!K)×1-B÷K+1
```

The latter expression is the basis of the definition of the estimating function

```
EST:(α*ω)÷(!ω)×1-α÷ω+1
```

Note that if *B*>0 then *B EST* ω is a decreasing function on the interval ω≥*B*, and that the values *B EST K* approach 0 as *K* is assigned larger and larger values. According to Theorem 6.1.2 and Identity 6.1.4:

```
(|(¯100↑K↓÷!⍳100)) POLY X)≤B EST K               [6.1.5]
```

for all arguments *X* for which *B*≥|*X*.

Consider the values for the function *EST* given in Table 6.1.6. Since 10 is the smallest value of *K* for which 0.0005>2 *EST K*, we assign *L*←10. Table 6.1.7 is an example of the tables of values of the polynomial with coefficient vector ÷!⍳100 to three decimal place accuracy.

K	2 *EST K*
2	6.0000
3	2.6667
4	1.1111
5	.4000
6	.1244
7	.0339
8	.0082
9	.0018
10	.0003
11	.0001

VALUES OF *EST*
TABLE 6.1.6

One of the important things to notice about this approximation procedure is that among the family of coefficient vectors ÷!⍳*N*, the fact that we were working with the coefficient vector ÷!⍳100 had very little to do with the final result. More precisely, for any integer *N* such that *N*>*K*, we have in place of Identity 6.1.3:

```
B×⌈/|K↓QT ÷!⍳N
B×⌈/|K↓÷1+⍳N-1
B÷K+1
```

X	3⍕(÷!⍳10) *POLY X*
0	1.000
0.1	1.105
0.2	1.221
0.3	1.350
0.4	1.492
0.5	1.649
0.6	1.822
0.7	2.014
0.8	2.226
0.9	2.460
1	2.718
1.1	3.004
1.2	3.320
1.3	3.669
1.4	4.055
1.5	4.482
1.6	4.953
1.7	5.474
1.8	6.050
1.9	6.686
2	7.389

APPROXIMATE VALUES OF (÷!⍳100) *POLY ω*
TABLE 6.1.7

Therefore *M* is *B*÷*K*+1 as before and in place of Inequality 6.1.5 we have

(|((-*N*)↑*K*↓÷!⍳*N*) *POLY X*)≤*B EST K*

for all arguments *X* such that *B*≥|*X*. Thus for arguments *X* such that 2≥|*X* and any integer *N* for which *N*>10, the polynomial with coefficient vector ÷!⍳10 is an approximation to the polynomial with coefficient vector ÷!⍳*N* which is accurate to three decimal places. For example, Table 6.1.7 is also a table of values of the polynomial with coefficient vector ÷!⍳1000 to three decimal place accuracy.

Another important (and perhaps obvious) point is that we can produce approximations of any desired accuracy. Consider Table 6.1.8, which extends Table 6.1.6. Evidently 17 is the smallest value of *K* for which 0.5*E*¯9>2 *EST K*. Thus for any argument *X* for which 2≥|*X*, the values of the polynomial with coefficient vector ÷!⍳17 are nine decimal place approximations to the values of the polynomial with coefficient vector ÷!⍳100, or ÷!⍳1000, and so on.

One final remark - for the family of coefficient vectors ÷!⍳*N*, the approximation procedure can be applied on any interval of arguments *X* of the form *B*≥|*X*. For example, it is obvious from Table 6.1.9 that 16 is the smallest value of *K* for which 0.0005>4 *EST K*. Therefore for arguments *X* for which 4≥|*X*, the values of the polynomial with coefficient vector ÷!⍳16 are three decimal place approximations to the values of the polynomial with coefficient vector ÷!⍳100, or ÷!⍳1000, and so on.

K	2 *EST K*
11	.0000615681
12	.0000101059
13	.0000015348
14	.0000002169
15	.0000000286
16	.0000000035
17	.0000000004
18	.0000000000
19	.0000000000
20	.0000000000

VALUES OF *EST*
TABLE 6.1.8

A EST K

		1	2	3	4	5	*A*
K	6	.0016	.1244	1.7719	13.2741	75.9549	
	7	.0002	.0339	.6943	6.5016	41.3360	
	8	.0000	.0082	.2441	2.9257	21.7983	
	9	.0000	.0018	.0775	1.2040	10.7646	
	10	.0000	.0003	.0224	.4541	4.9338	
	11	.0000	.0001	.0059	.1576	2.0970	
	12	.0000	.0000	.0014	.0506	.8282	
	13	.0000	.0000	.0003	.1051	.3049	
	14	.0000	.0000	.0001	.0042	.1050	
	15	.0000	.0000	.0000	.0011	.0339	
	16	.0000	.0000	.0000	.0003	.0103	
	17	.0000	.0000	.0000	.0001	.0030	
	18	.0000	.0000	.0000	.0000	.0008	
	19	.0000	.0000	.0000	.0000	.0002	
	20	.0000	.0000	.0000	.0000	.0001	

VALUES OF *EST*
TABLE 6.1.9

We can summarize the results discussed in the preceding paragraphs in terms of the inequality:

(*B EST N*⌊*M*)≥|((÷!⍳*N*) *POLY X*)-(÷!⍳*M*) *POLY X* [6.1.10]

for non-negative integers *N* and *M* such that *B*≤*N*⌊*M* and arguments *X* such that *B*≥|*X*. Since the function values *B EST K* are small when *K* is large relative to *B*, it follows from the above inequality that for arguments *X* such that *B*≥|*X* the values of the polynomials with coefficient vectors ÷!⍳*N* and ÷!⍳*M* are close to one another for integers *N* and *M* which are large relative to *B*. This suggests that the family of polynomials with coefficient vectors ÷!⍳*N* can be used to define a (possibly) new function.

Specifically, we implicitly define the monadic scalar function to be called *E* by defining each function value *E X* to be the value approached by the function values `(÷!⍳N) POLY X` as *N* is assigned large values. To see that such a value *E X* does indeed exist, we need only show how to determine any specified number of digits in the decimal representation of *E X*. For example, if we wish to know the values of *E* to 15 decimal place accuracy on the interval `100≥|X`, we determine an integer *K* for which `0.5E¯15≥100 EST K`. Then for arguments *X* such that `100≥|X` and any two integers *N* and *M* for which `K≤N⌊M`, Inequality 6.1.10 shows that the values of the polynomials with coefficient vectors `÷!⍳N` and `÷!⍳M` agree to 15 decimal digits. Evidently the first 15 decimal digits of the value *E X* must be the same as those of the values of any one of these polynomials.

The function *E* will be studied extensively in this text, beginning in the next section.

The proof of Theorem 6.1.2 is as follows. For a vector *C* and a scalar *X* the vector defined by the expression `C×X*⍳⍴C` is called the <u>vector of terms</u> of the polynomial with coefficient vector *C*. For an integer *K* the scalar defined by `+/K↑C×X*⍳⍴C` is called a <u>prefix</u> of the polynomial with coefficient vector *C* and the scalar defined by `+/K↓C×X*⍳⍴C` is called a <u>suffix</u>. Evidently

```
+/C×X*⍳⍴C
(+/K↑C×X*⍳⍴C)++/K↓C×X*⍳⍴C
```

and moreover

```
+/K↓C×X*⍳⍴C
((-⍴C)↑K↓C) POLY X
```

Thus to prove the theorem we must estimate the values of the suffix `+/K↓C×X*⍳⍴C`.

The central idea of the proof is to compare the values of `+/K↓C×X*⍳⍴C` with the values of polynomials whose coefficient vectors conform to a particularly simple pattern. Namely, the polynomials used for comparison are scalar multiples of polynomials whose coefficient vectors are of the form `N⍴1` (Inequality 6.1.12 below). For appropriate scalars *M*, the values of all such polynomials are easy to estimate (Inequality 6.1.15 below).

The proof of the theorem begins with the inequality

```
(|+/K↓C×X*⍳⍴C)≤+/K↓(|C)×(|X)*⍳⍴C
```

and if `B≥|X`:

```
(+/K↓(|C)×(|X)*⍳⍴C)≤+/K↓|C×B*⍳⍴C
```

Combining these two inequalities yields for scalars *X* such that *B*≥|*X*:

```
(|+/K↓C×X*⍳⍴C)≤+/K↓|C×B*⍳⍴C                        [6.1.11]
```

Next, we have from Exercise 6.3a:

```
QT K↓|C×B*⍳⍴C
B×QT K↓|C
```

Thus if the scalar *M* is defined as in Theorem 6.1.2, that is

```
M←B×⌈/QT K↓|C
```

then

```
∧/(QT K↓|C×B*⍳⍴C)≤M
```

Consequently, according to 6.1.1 and Exercise 6.3b:

```
∧/(K↓|C×B*⍳⍴C)≤(|C[K]×B*K)×M*⍳(⍴C)-K
```

and therefore:

```
(+/K↓|C×B*⍳⍴C)≤(|C[K]×B*K)×+/M*⍳(⍴C)-K          [6.1.12]
```

For the geometric progression defined by the expression *M**⍳*N* we have the identity

```
+/M*⍳N                                              [6.1.13]
(1-M*N)÷1-M
```

If (0≤*M*)∧*M*<1, it follows that (0≤*M***N*)∧1>*M***N* and consequently

```
(0<1-M*N)∧1≥1-M*N                                   [6.1.14]
```

Combining 6.1.13 and 6.1.14 yields:

```
(+/M*⍳N)≤÷1-M                                       [6.1.15]
```

Finally, combining this inequality with 6.1.11 and 6.1.12 yields:

```
(|+/K↓C×X*⍳⍴C)≤(|C[K]×B*K)÷1-M
```

which, for scalars *X*, is equivalent to the inequality in the statement of Theorem 6.1.2.

Other applications of Theorem 6.1.2 will be discussed in the exercises. ⍠4

6.2 GROWTH AND DECAY FUNCTIONS

The monadic scalar function *GD* is called a <u>growth function</u> or a <u>decay function</u> if there exists a scalar *R* such that for *GD* and its derivative *D̲GD*:

```
D̲GD X                                              [6.2.1]
R×GD X
```

In words, the derivative of *GD* is identical to a constant multiple of *GD* itself, or *GD* is proportional to its own derivative. A function *GD* for which the Identity 6.2.1 is valid is called a growth function if it is increasing and a decay function if it is decreasing.

It is not difficult to see that no polynomial can be a growth or decay function. If there was such a polynomial, say with coefficient vector *C*, and if *C* had no trailing zeros (that is, if 0≠¯1↑*C*), then Identity 6.2.1 would imply the identity

```
DERIV C
R×C
```

and consequently ρ*DERIV C*
R×*X* ρ*C*. However,
(ρ*DERIV C*)=¯1+ρ*C*.

Even though no polynomial can be a growth function, it is not difficult to construct vectors *C* for which

```
DERIV C                                            [6.2.2]
¯1↓R×C
```

According to the definition of the function *DERIV*:

```
DERIV C
1↓C×ιρC
```

and consequently 6.2.2 is equivalent to

```
1↓C×ιρC
        ¯1↓R×C
```

This identity can be written element-by-element as follows:

```
(R×C[0])=C[1]
(R×C[1])=C[2]×2
(R×C[2])=C[3]×3
```

and in general:

```
(R×C[I])=C[I+1]×I+1
```

Evidently if the element *C*[0] is known, then the first equation above can be solved for the element . The second equation can then be solved for the element *C*[2], the third can then be solved for *C*[3], and so on. If the scalar *A* represents the element *C*[0], that is if *A*=*C*[0], the result of this solution procedure is the vector

```
A××\1,R÷1+⍳N                                 [6.2.3]
```

or equivalently

```
A×(R*⍳N+1)÷!⍳N+1                                        ⌹22
```

For example, if `R←2` and `C←3×(R*⍳10)÷!⍳10`, then `DERIV C` is `¯1↓R×C`. Note that

```
(A×(R*⍳N)÷!⍳N)POLY ω                         [6.2.4]
A×(÷!⍳N)POLY R×ω
```

so that in analyzing the polynomials with coefficient vectors defined by the expression 6.2.3 it is sufficient to analyze the polynomials with coefficient vectors `÷!⍳N`. ⌹23

Explicit, approximate definitions of growth and decay functions will be discussed in the exercises. The construction is based on the formal implicit definition of these functions, which is now quite easy to describe. As in Section 6.1, the function called *E* is implicitly defined by defining each function value `E X` to be the value approached by the function values `(÷!⍳N) POLY X` as *N* is assigned large values. It is then true that the derivative `D̲E X` is identical to the value approached by the function values `(DERIV ÷!⍳N) POLY X` as *N* is assigned large values. (The proof of this fact is a topic for more advanced texts.) Since:

```
DERIV ÷!⍳N
÷!⍳N-1
```

it follows that the values `(DERIV ÷!⍳N) POLY X` also approach the value `E X`. Consequently:

```
D̲E X
E X
```

In words, the function *E* is identical to its own derivative.

In general, using Identity 6.2.4 as a guide, for scalars *A* and *R* we define the function called *GD* as

```
GD:A×E R×ω                                   [6.2.5]
```

Then *GD* is either a growth function or a decay function. For *GD* and its derivative function *D̲GD* we have the following sequence of identities:

```
D̲GD ω
A×R×D̲E R×ω
A×R×E R×ω
R×A×E R×ω
R×GD ω
```

In words, the function *GD* is a constant multiple of its own derivative.

Since every growth or decay function is defined in terms of the function *E* as in 6.2.5, there is no need for the name *GD* and we simply refer to the growth or decay function defined by the expression *A*×*E R*×ω.

It is important to realize that in the above construction we did not assume that growth and decay functions existed. Instead, Identity 6.2.1 and Theorem 6.1.2 were used to formulate an implicit definition of these functions. In most cases of practical interest solutions of differential equations such as Identity 6.2.1 represent natural phenomena, and therefore their existence can be assumed. It is, then, often possible to produce coefficient vectors of Taylor polynomials of the solutions directly from the differential equations. For example, assume that there is a function *E* ω which is identical to its derivative. Then, differentiating both expressions in the identity

```
D̲E ω
E ω
```

yields the fact that D̲D̲*E* ω equals D̲*E* ω, and consequently D̲D̲*E* ω equals *E* ω. Thus the second derivative of *E* is also identical to *E*. Evidently this argument can be continued to show that every higher derivative of *E* is identical to *E*. Consequently the Taylor coefficient function of *E* is

```
T̲E:(E ω)÷!⍳α+1
```

In particular, if *E* 0 is 1, the coefficient vector of the (*N*-1)th order Taylor polynomial of *E* at 0 is 1÷!⍳*N*, or equivalently ÷!⍳*N*.

The simple form of the Taylor coefficient function T̲*E* leads directly to the following addition formula for *E* ω:

```
E ω+α
(E α)×E ω
```

The proof of this identity will be discussed in the exercises. Comparing it with the left side of Identity 4.1.1 suggests a relationship between *E* ω and the power function. In fact, such a relationship exists, and is described by the following identity:

```
E ω
(E 1)*ω
```

The value *E* 1 is a famous transcendental number which is commonly denoted by lower case e. Approximate values of *E* 1 are produced by the expression (÷!⍳*N*) *POLY* 1, or equivalently +/÷!⍳*N*.

Growth and Decay Functions; Another Approach. We will now pursue the construction of growth and decay functions from the point of view of vectors of approximate values. Specifically, for a scalar *R* and a vector of arguments *X* we will construct vectors *Y* for which

```
      (DIFF Y)÷DIFF X                                    [6.2.6]
      R×¯1↓Y
```

The following sequence shows that 6.2.6 and Identity 6.2.7 below are equivalent.

```
      ((DIFF Y)÷DIFF X)=R×¯1↓Y
      (DIFF Y)=(R×¯1↓Y)×DIFF X
      ((1↓Y)-¯1↓Y)=(¯1↓Y)×R×DIFF X
      (1↓Y)=(¯1↓Y)×1+R×DIFF

      1↓Y                                                [6.2.7]
      (¯1↓Y)×1+R×DIFF X
```

Identity 6.2.7 can be written element-by-element as follows:

```
      Y[1]=Y[0]×1+R×(DIFF X)[0]
      Y[2]=Y[1]×1+R×(DIFF X)[1]
```

and in general:

```
      Y[I+1]=Y[I]×1+R×(DIFF X)[I]
```

As before, if the element `Y[0]` is known then these equations can be solved in succession for the elements `Y[1]`, `Y[2]`, and so on. If the scalar named `Y0` represents the first element of the vector *Y*, then the result of this solution procedure is the vector

```
      Y0××\1,1+R×DIFF X                                  [6.2.8]
```

⌹40

Thus Identity 6.2.6 is valid for a vector *X* and the vector *Y* defined as in 6.2.8. For example:

```
      R←2
      X←0 1 2.5 3 5 8

      ⎕←Y←3××\1,1+R×DIFF X
3 9 36 72 360 2520

      ∧/(R×¯1↓Y)=(DIFF Y)÷DIFF X
1
```

In particular, let *R* be `1` and `Y0` be `E X[0]`. Using 6.2.8 as a guide, for a vector of arguments *X* define the vector *Y* as follows:

```
      Y←(E X[0])××\1,1+DIFF X
```

Then according to Identity 6.2.6:

```
¯1↓Y                                                    [6.2.9]
(DIFF Y)÷DIFF X
```

Roughly speaking, for vectors *X* for which the values of the scalar defined by ⌈/|*DIFF X* are close to 0, Identity 6.2.9 is "approximately equivalent" to the fact that *E* is equivalent to its derivative *DE*. One might then suspect that the vector *Y* is approximately equal to the vector of values *E X*, and in fact it is. For example, according to the Table 6.1.9, the values of the polynomial with coefficient vector ÷!⍳7 are three decimal digit approximations of the values of the function *E* for arguments *X* such that 1≥|*X*. Define the vectors *X* and *Y* as follows:

```
X←0.5+0.01×⍳6
Y←×\((÷!⍳7) POLY X[0])××\1,1+DIFF X
```

(note that the approximate value (÷!⍳7) *POLY X*[0] has been used for *E X*[0]) and compare the vectors *Y* and (÷!⍳7) *POLY X*:

```
      4⍕Y
1.6487 1.6652 1.6819 1.6987 1.7157 1.7328

      4⍕(÷!⍳7) POLY X
1.6487 1.6653 1.6820 1.6989 1.7160 1.7332
```
⊞41

The above procedure for producing vectors of approximate values of a solution of a differential equation can be extended to methods that are similar to those of Section 5.5 for producing approximate values of definite integrals. Detailed discussions of the extensions can be found in most texts on numerical analysis. We will not pursue these methods here, but instead will concentrate on Taylor polynomials.

Primary Growth and Decay Functions. The primary function ⋆⍵ is defined by the identity

```
⋆⍵
E ⍵
```

The family of functions defined by *A*×⋆*R*×⍵ is called the family of growth and decay functions.

The derivative of ⋆⍵ is ⋆⍵, and more generally the derivative of ⋆*R*×⍵ is the function *R*×⋆*R*×⍵. The addition formula for ⋆⍵ is

```
⋆⍵+⍺
(⋆⍵)×⋆⍺
```

and the Taylor coefficient function is

    TE:(*ω)÷!⍳α+1   ⌹51

<u>Electrical</u> <u>Circuits</u>. In the electrical circuit in Figure 6.2.10, the component labelled *R* is a resistor of *R* ohms, that labelled *L* is an inductance of *L* henries, that labelled *C* is a capacitor of *C* farads, and the one labelled *EF* is a source of electromotive force of *EF* volts.

The functions to be called *Q* and *I* represent the quantity of electricity or <u>charge</u> in the capacitor and the rate of flow of electricity or <u>current</u> into the capacitor respectively. That is, at <u>time</u> *T*, the charge in the capacitor is *Q T* coulombs and the current into the capacitor is *I T* amperes. By definition of the functions *Q* and *I*:

    DQ T   [6.2.11]
    I T

The resistance *R* is a measure of the electrical energy lost as heat when current flows through the resistor; the unit of resistance is chosen so that *R*×*I T* is identical to the voltage drop across the resistor.

The inductance *L* is a measure of the resistance of the coil to a change in current; the unit of inductance is chosen so that *L*×*DI T* is identical to the voltage drop across the coil, where *DI* is the derivative of the function *I*. Note that since *I ω* is *DQ ω* (Identity 6.2.11), *DI ω* is *DDQ ω*, where *DDQ* is the second derivative of *Q*. Thus the voltage drop across the coil is also expressed by *L*×*DDQ T*.

The capacitance *C* is a measure of the resistance of the capacitor to a change in voltage; the unit of capacitance is chosen so that (÷*C*)×*Q T* is identical to the voltage drop across the capacitor.

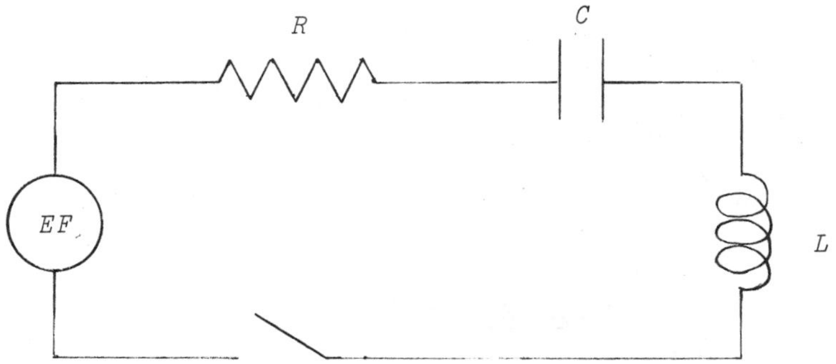

AN ELECTRIC CIRCUIT
FIGURE 6.2.10

According to Kirchhoff's law, when the switch in the circuit is closed the voltage *EF* at the source is identical to the sum of the voltage drops across the resistor, the coil, and the capacitor. Consequently:

```
EF                                        [6.2.12]
(L×D̲I T)+(R×I T)+(÷C)×Q T
```

or equivalently:

```
EF                                        [6.2.13]
(L×D̲D̲Q T)+(R×D̲Q T)+(÷C)×Q T
```

If the coil is not present in the circuit illustrated above then `L=0` and Identity 6.2.13 becomes

```
EF                                        [6.2.14]
(R×D̲Q T)+(÷C)×Q T
```

Functions *Q* for which Identity 6.2.14 is valid can be constructed in the same way that growth and decay functions were constructed. However, for this particular example it is not necessary to repeat that construction; the functions we seek can be defined in terms of the already constructed growth and decay functions. Suppose that Identity 6.2.14 is valid for the function called *Q* and define the function `Q0:(-EF×C)+Q ω`. Then for the derivatives `D̲Q` and `D̲Q0` of *Q* and `Q0` respectively we have

```
D̲Q T
D̲Q0 T
```

Substituting `(EF×C)+Q0 T` for `Q T` and `D̲Q0 T` for `D̲Q T` in Identity 6.2.14 yields the sequence of identities:

```
EF
(R×D̲Q0 T)+(÷C)×(EF×C)×Q0 T
EF+(R×D̲Q0 T)+(÷C)×Q0 T
```

Consequently:

```
0
(R×D̲Q0 T)+(÷C)×Q0 T
```

or equivalently

```
D̲Q0 T
(-÷R×C)×Q0 T
```

Evidently `Q0` is a decay function, so that there exists a scalar *A* for which `Q0:A⋆-ω÷R×C`. Consequently:

```
Q:(EF×C)+A××-ω÷R×C
```

It is not difficult to check that for any scalar A, Identity
6.2.14 is valid for this function Q.   The scalar A can be
assigned a value if there is information available which
tells us one value of the function Q.   For example, we may
assume that the time $T \leftarrow 0$ is chosen to be the instant at
which the switch is closed. At that instant there is no
charge in the capacitor, which means that $0 = Q\ 0$.   To
determine the value to be assigned to A so that $0 = Q\ 0$,
consider the sequence of identities:

        0
        $Q\ 0$
        $(EF \times C) + A \times \star 0$
        $(EF \times C) + A$

Evidently we assign A to be $-EF \times C$ and therefore the function
Q such that $0 = Q\ 0$ and for which Identity 6.2.14 is valid is

        $Q : (EF \times C) \times 1 - \star - \omega \div R \times C$ [6.2.15]

        Solutions of the more general differential equation
6.2.13 can be defined in terms of the growth function $E\ \omega$
and the circular functions to be introduced in the next
section.   Detailed discussions of solutions of identities of
the form 6.2.13 can be found in most elementary texts on
differential equations.   ⊟54

6.3   THE CIRCULAR FUNCTIONS

        Figure 6.3.1 is a circle of radius 1 centered at the
origin and the scalar A represents the indicated angle.   The
angle is measured in degrees.   The coordinate point which is

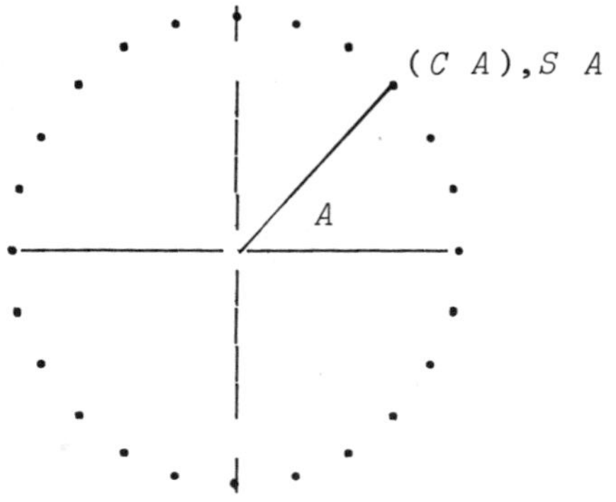

CIRCULAR FUNCTIONS   ANGLE MEASUREMENT
    FIGURE 6.3.1   FIGURE 6.3.2

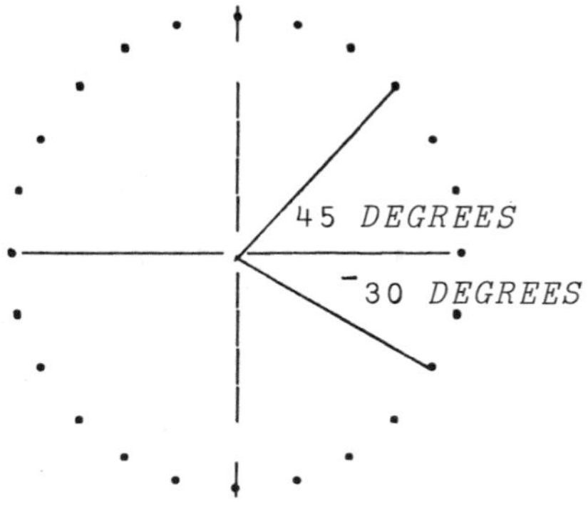

the intersection of the line segment forming one side of the angle *A* and the circle is represented by the values of a pair of functions *C* and *S*. That is, the coordinate point is (*C A*),*S A*. The functions *C* and *S* are called <u>circular functions</u> for obvious reasons. As usual, if *A*>*0* then the angle represented by *A* is measured in the counterclockwise direction from the positive horizontal axis, while if *A*<*0* the angle is measured in the clockwise direction (Figure 6.3.2). Also, each vector of the form (*C A*),*S A* is a direction vector (Section 2.5) and conversely, for each direction vector *V* there is an argument *A* for which *V*[0]=*C A* and *V*[1]=*S A*.

It is not difficult to make tables of values for the circular functions. For that purpose we will use the functions *NORM*:((ω⋆2)+⌽ω⋆2)⋆0.5 and *DIRECTION*:ω÷*NORM* ω from Section 2.5. Note that this definition of *NORM* is different from the one in Section 2.5. The point is that *DIRECTION* now applies along the rows of a matrix *M* and produces a matrix of the same size whose rows are the direction vectors of the rows of *M*. To make tables of values of the circular functions we begin with a few obvious values (Figure 6.3.3). These values are arranged in the table *T* as follows:

```
      4⍕T
   .0000  1.0000   .0000
 45.0000   .7071   .7071
 90.0000   .0000  1.0000
```

The elements of the first column of *T* represent the angle *A* and the elements of the second and third columns represent the function values *C A* and *S A* respectively. Now form the table *R*, whose rows are the pairwise averages of successive

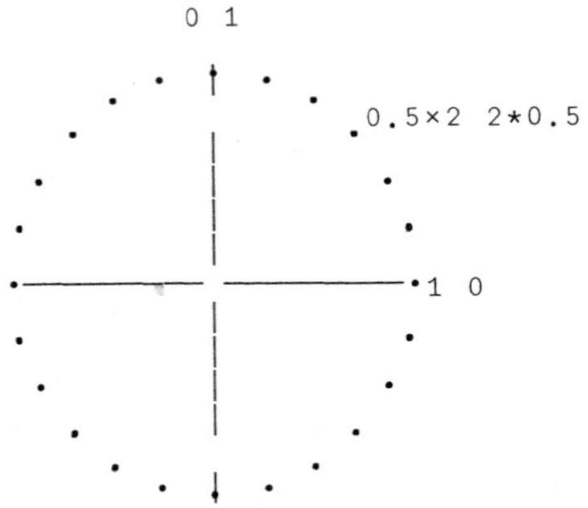

SOME VALUES OF THE CIRCULAR FUNCTIONS

FIGURE 6.3.3

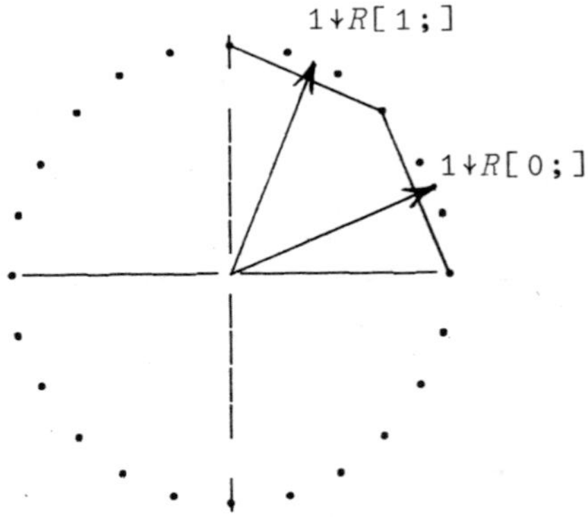

TWO NEW VALUES OF THE CIRCULAR FUNCTIONS

FIGURE 6.3.4

rows of *T*:

```
      R←0.5×(1 0↓T)+¯1 0↓T
      4⍕R
 22.5000   .8536   .3536
 67.5000   .3536   .8536
```

For each index *I* the vector 1↓R[I;] represents the midpoint of the line segment joining the points 1↓T[I;] and 1↓T[I+1;] (Section 2.5). The element R[I;0] represents the angle formed by the horizontal axis and the line segment joining the origin to the point 1↓R[I;] (Figure 6.3.4).

The rows of the table *DIRECTION* 0 1↓R are the direction vectors of the rows of 0 1↓R. Evidently the elements of the row *I* of *DIRECTION* 0 1↓R are the values of *C* and *S* for the argument R[I;0]. Replace the last two columns of the table *R* with the table of direction vectors:

```
      R[;1 2]←DIRECTION R[;1 2]
      4⍕R
 22.5000   .9239   .3827
 67.5000   .3827   .9239
```

Next weave *R* into *T* to form an extended table of values of *C* and *S*:

```
      Q←T,[0] R
      4⍕Q
   .0000  1.0000   .0000
 45.0000   .7071   .7071
 90.0000   .0000  1.0000
 22.5000   .9239   .3827
                 67.5000   .3827   .9239

      ⎕←P←⍋Q[;0]                         T←Q[P;]
0 3 1 4 2
```

(*P* is the permutation for which the elements of the vector Q[P;0] are in increasing order.)

```
      4⍕T
   .0000  1.0000   .0000
 22.5000   .9239   .3827
 45.0000   .7071   .7071
 67.5000   .3827   .9239
 90.0000   .0000  1.0000
```

If we repeat this procedure four new values of the circular functions are produced.

```
      R←0.5×(1 0↓T)+¯1 0↓T
      R[;1 2]←DIRECTION R[;1 2]
      Q←T,[0] R
      P←⍋Q[;0]
      T←Q[P;]
```

Evidently this procedure can be continued to produce extensive tables of the circular functions *C* and *S*.

```
     4⍕T
 .0000  1.0000   .0000
11.2500   .9808   .1951
22.5000   .9239   .3827
33.7500   .8315   .5556
45.0000   .7071   .7071
56.2500   .5556   .8315
67.5000   .3827   .9239
78.7500   .1951   .9808
90.0000   .0000  1.0000
```

VALUES OF THE CIRCULAR FUNCTIONS
TABLE 6.3.5

Circular functions can also be defined in terms of units other than degrees. For a scalar *X* mark off an arc of length |*X* on the perimeter of the unit circle starting at the point 1 0 and proceeding counterclockwise if *X*>0 or clockwise if *X*<0. The coordinates of the endpoint of the arc are represented by a pair of functions of *X* called *COS* (for cosine) and *SIN* (for sine), as in Figure 6.3.6.

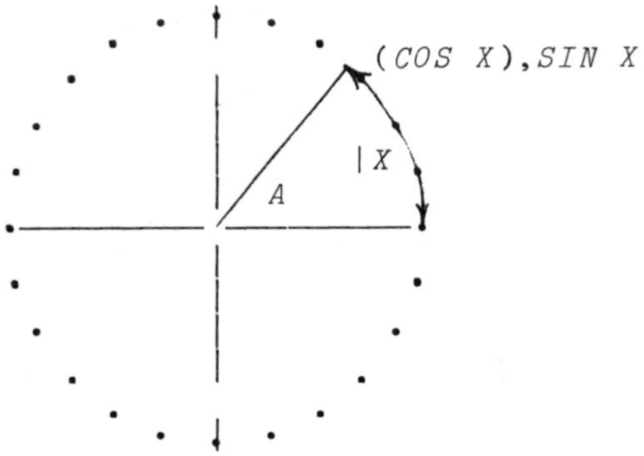

CIRCULAR FUNCTIONS
FIGURE 6.3.6

There is a simple relation between the scalar *X* and the angle *A*, illustrated in Figure 6.3.6. A full sweep around the circle produces an angle of 360 degrees and also an arclength of ○2 (the circumference of the circle; the monadic circle function ○ω is equivalent to *PI*×ω, where *PI* is the circumference of a circle of diameter 1). Evidently we have the simple proportionality relation (*R*÷360)=*X*÷○2 and therefore *X*=*A*×○÷180. In particular

```
C A                     S A
COS A×○÷180             SIN A×○÷180
```

Consequently if we replace the column of arguments *T*[;0] of a table of values of *C* and *S* with *T*[;0]×○÷180, we will have a table of values for *COS* and *SIN*.

```
      T[;0]←T[;0]×○÷180
              4⍕T
           .0000 1.0000  .0000
           .1963  .9808  .1951
           .3927  .9239  .3827
           .5890  .8315  .5556
           .7854  .7071  .7071
           .9817  .5556  .8315
          1.1781  .3827  .9239
          1.3744  .1951  .9808
          1.5708  .0000 1.0000
```

VALUES OF *SIN* and *COS*
TABLE 6.3.7

The functions *SIN* and *COS* can also be constructed as solutions of a differential equation, in much the same way as the growth function *⋆ω*. The details will be developed in the exercises. ⌹56

The Addition Formula for *COS*.

```
      COS X+Y
      ((COS X)×COS Y)-(SIN X)×SIN Y
```

PROOF. (See Figure 6.3.8). Suppose that *P* represents the point (*COS X+Y*),*SIN X+Y* and *Q* the point (*COS Y*),*SIN Y* and *R* the point (*COS X*),-*SIN X*. The length of arc from *P* to the point 1 0 is *X+Y*, as is the length of arc from *Q* to *R*. Thus the distance from *P* to the point 1 0 is identical to the distance from *Q* to *R*, and the squares of these

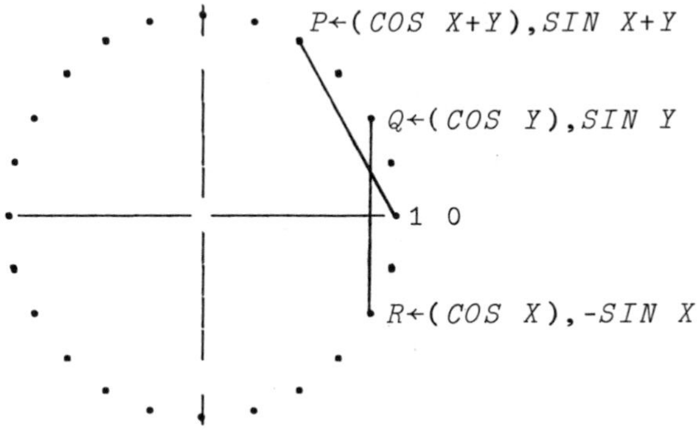

ADDITION FORMULA FOR *COS ω*
FIGURE 6.3.8

distances are also identical. That is:

```
((¯1+COS X+Y)⋆2)+(SIN X+Y)⋆2                          [6.3.9]
(((COS Y)-COS X)⋆2)+((SIN Y)--SIN X)⋆2
```

For the first expression in this identity we have:

```
((¯1+COS X+Y)⋆2)+(SIN X+Y)⋆2                          [6.3.10]
1+(¯2×COS X+Y)+((COS X+Y)⋆2)+(SIN X+Y)⋆2
2-2×COS X+Y
```

For the second expression in 6.3.9 we have:

```
(((COS Y)-COS X)⋆2)+((SIN Y)--SIN X)⋆2                [6.3.11]
2-2×((COS X)×COS Y)-(SIN X)×SIN Y
```

According to 6.3.9, the first expression in 6.3.10 is equivalent to the first expression in 6.3.11. Therefore the last expression in 6.3.10 is equivalent to the last expression in 6.3.11 and

```
2-2×COS X+Y
2-2×((COS X)×COS Y)-(SIN X)×SIN Y
```

Therefore, as stated:

```
COS X+Y
((COS X)×COS Y)-(SIN X)×SIN Y
```

The Addition Formula for *SIN*.

```
SIN X+Y
((SIN X)×COS Y)+(COS X)×SIN Y
```

PROOF. (see Figure 6.3.12). Suppose that *P* represents the point (*COS X+Y*),*SIN X+Y* and *Q* the point (*COS Y*),*SIN Y* and *R* the point (*SIN X*),*COS X*. The length of arc from *P* to the point 0 1 is identical to the length of arc from *R* to *Q*,

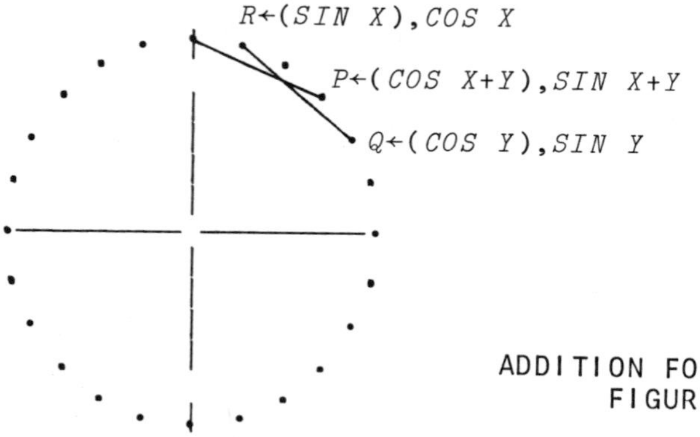

ADDITION FORMULA FOR *SIN ω*
FIGURE 6.3.12

and consequently the square of the distance from *P* to the point `0 1` is identical to the square of the distance from *R* to *Q*. That is:

```
((COS X+Y)*2)+(¯1+SIN X+Y)*2                      [6.3.13]
(((COS Y)-SIN X)*2)+((SIN Y)-COS X)*2
```

For the first expression in 6.3.13 we have:

```
((COS X+Y)*2)+(¯1+SIN X+Y)*2                      [6.3.14]
2-2×SIN X+Y
```

For the last expression in 6.3.13:

```
(((COS Y)-SIN X)*2)+((SIN Y)-COS X)*2             [6.3.15]
2-2×((SIN X)×COS Y)+(COS X)×SIN Y
```

Therefore, since the last expressions in 6.3.14 and 6.3.15 agree:

```
SIN X+Y
((SIN X)×COS Y)+(COS X)×SIN Y
```
⍠68

<u>The Derivatives of the Circular Functions</u>. In order to determine the cline functions and derivatives of the circular functions we must write their addition formulas in normal form (Section 1.4). For that purpose consider the functions `(SIN α)÷α` and `(¯1+COS α)÷α`. The values of both functions for the argument `0` are the same as `0÷0` and are therefore indeterminate. However, the functions can be completed at the argument `0` in the manner in which the difference-quotient function was completed to produce the cline function in Section 1.4. That is, as we will prove below, there are constants *A* and *B* for which the values `(SIN X)÷X` and `(¯1+COS X)÷X` approach *A* and *B* as *X* approaches `0`. Thus we define the functions

```
DC:(¯1+COS α)÷α : 0=α : B
DS:(SIN α)÷α : 0=α : A
```

and

```
C̲COS:((DC α)×COS ω)-(DS α)×SIN ω
C̲SIN:((DC α)×SIN ω)+(DS α)×COS ω
```

Then according to the addition formulas for *SIN* and *COS*:

```
COS ω+α                                           [6.3.16]
(COS ω)+α×α C̲COS ω
```

and

```
SIN ω+α                                           [6.3.17]
(SIN ω)+α×α C̲SIN ω
```
⍠70

Thus *C̲COS* and *C̲SIN* are the cline functions of *COS* and *SIN* respectively. Consequently, the derivative of *SIN ω* is

0 *CSIN ω* and the derivative of *COS ω* is 0 *CCOS ω*, or equivalently:

DCOS ω [6.3.18]
(*B*×*COS ω*)-*A*×*SIN ω*

DSIN ω [6.3.19]
(*B*×*SIN ω*)+*A*×*COS ω*

We will also show that *A* is 1 and *B* is 0, and consequently:

DCOS ω *DSIN ω* [6.3.20]
-*SIN ω* *COS ω*

We will first prove that *A* is 1. For that purpose consider Figure 6.3.21. The arc in the diagram lies on the circle of radius 1 centered at the origin, so that *P* represents the point (*COS Y*),*SIN Y* for some scalar *Y*. The line through the point *Q* and the origin bisects the arc so that *Q* represents the point (*COS* 0.5×*Y*),*SIN* 0.5×*Y*. The point represented by *R* lies on this line and on the vertical line rising from the point 1 0. It was pointed out in Section 2.5 that since the point represented by *R* lies on the line through the origin and the point (*COS* 0.5×*Y*),*SIN* 0.5×*Y*, there must be a scalar *V* for which the coordinates of this point are (*V*×*COS* 0.5×*Y*),*V*×*SIN* 0.5×*Y*. On the other hand, *R* lies on the vertical line rising from the point 1 0 and so the first coordinate of this point must be 1. Consequently:

1=*V*×*COS* 0.5×*Y* and therefore *V*=÷*COS* 0.5×*Y*

Therefore the point represented by *R* is the coordinate point 1,(*SIN* 0.5×*Y*)÷*COS* 0.5×*Y*.

The area of the triangle with vertices at the points *R* and 1 0 and the origin is 0.5×(*SIN* 0.5×*Y*)÷*COS* 0.5×*Y*.

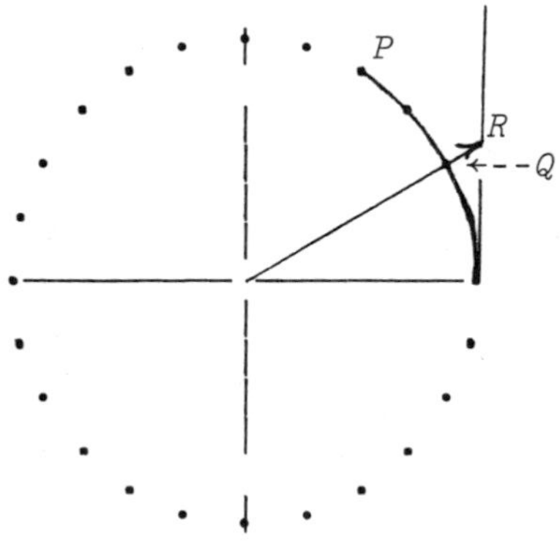

USED IN·THE PROOF OF 6.3.20
FIGURE 6.3.21

The area of the sector of the unit circle with vertices at the points Q and `1 0` and the origin is `0.5` times the length of arc between `1 0` and Q. Since this length of arc is `0.5×Y`, the area of this sector is `0.25×Y`. Since the area of this sector is clearly less than the area of the triangle with vertices at R, `1 0` and the origin we have the inequality

    (0.25×Y)<0.5×(SIN 0.5×Y)÷COS 0.5×Y

or equivalently

    (COS 0.5×Y)<(SIN 0.5×Y)÷0.5×Y   [6.3.22]

Next, Y, which is the length of arc between P and the point `1 0`, is clearly greater than the distance from P to the point `1 0`, so that `Y*2` is greater than the square of this distance. That is:

    (Y*2)>((1-COS Y)*2)+(SIN Y)*2

or equivalently

    (1-0.5×Y*2)<COS Y   [6.3.23]

Finally, Y is greater than the height of the point P above the horizontal axis, which is `SIN Y`. That is:

    Y>SIN Y

or equivalently

    1>(SIN Y)÷Y   [6.3.24]

Now substitute `2×S` for Y in 6.3.22 and S for Y in 6.3.23 and 6.3.24. Three inequalities result:

    (COS S)<(SIN S)÷S
    (1-0.5×S*2)<COS S
    1>(SIN S)÷S

Combining the first two of these inequalities into one yields the pair of inequalities:

    (1-0.5×S*2)<(SIN S)÷S   and   1>(SIN S)÷S

As S is assigned values close to `0` the expression `1-0.5×S*2` assumes values close to `1`. Since the values of the expression `(SIN S)÷S` are greater than these values and at the same time smaller than `1` it follows that the expression `(SIN S)÷S` assumes values close to `1` as S is assigned values close to `0`. Thus A is `1`.

The fact that B is `0` follows directly from the fact that A is `1`. The proof will be developed in the exercises. ⌹71

<u>Primary</u> <u>Circular</u> <u>Functions</u>.   The primary sine and cosine functions are denoted by 1○ω and 2○ω respectively and defined by the identities:

    1○ω   2○ω
    SIN ω   *COS* ω

The derivatives of 1○ω and 2○ω are 2○ω and -1○ω respectively, and their second derivatives are -1○ω and -2○ω respectively.   More generally, the derivatives of the functions defined by 1○R×ω and 2○R×ω are R×2○R×ω and -R×1○R×ω respectively, and their second derivatives are -(R×2)×1○R×ω and -(R*2)×2○R×ω respectively.   For the functions 1○ω and 2○ω we have the addition formulas

    2○ω+α   1○ω+α
    -/×/2 1○.○ω,α   +/×/1 0⊖2 1○.○ω,α

The Taylor coefficient functions of 1○ω and 2○ω are

         *T̲C*1:((α+1)ρ1 1 ¯1 ¯1×1 2 1 2○ω)÷!ια+1
    and
         *T̲C*2:((α+1)ρ1 ¯1 ¯1 1×2 1 2 1○ω)÷!ια+1

respectively.   ⊟72

<u>Mechanical</u> <u>Vibrations</u>.   Consider a weight suspended from a spring attached to a beam, as illustrated in Figure 6.3.25. We assume that the weight is at its equilibrium position, that is, it is not moving.   According to Hooke's law, if the weight is set in vertical motion and if its displacement from the equilibrium position is small, then the acceleration imparted to the weight by the tension in the spring is proportional to the displacement of the weight from its equilibrium position.   In order to state Hooke's law mathematically we assume that the spring and the weight are represented in a coordinate system such as the one in Figure 6.3.25.

     Suppose that the function called *POS* represents the position of the vibrating spring relative to its equilibrium position.   Precisely, at time *T* the function value *POS T* is the coordinate of the center of the weight.   Thus if 0<*POS T* then the weight is below the equilibrium position and if 0>*POS T* it is above.   The second derivative *D̲D̲POS T* is identical to the acceleration of the weight at time *T*.   Thus according to Hooke's law, *D̲D̲POS T* is proportional to *POS T*.

     Evidently the total force on the spring always pulls the weight towards the equilibrium position.   In particular, if 0<*POS T* then the spring pulls the weight in the negative direction, which means that the acceleration is negative. That is, 0>*D̲D̲POS T* if 0<*POS T*.   Similarly, 0<*D̲D̲POS T* if

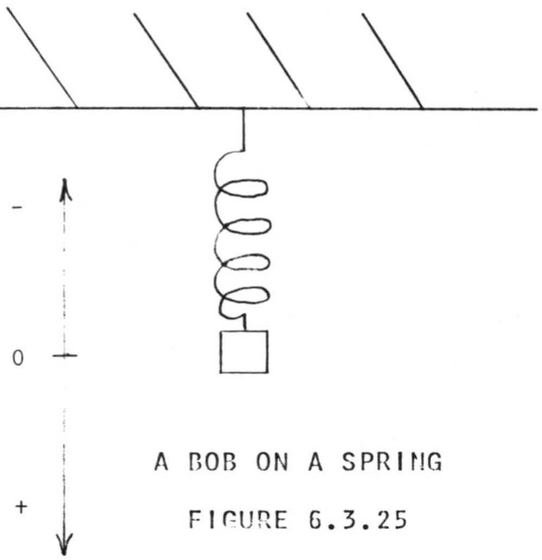

A BOB ON A SPRING

FIGURE 6.3.25

0>*POS T*. Consequently Hooke's law is the identity

```
DDPOS T
-D×POS T
```

where *D* is a constant for which 0<*D*. As is usual in the vibrating spring problem we will substitute *K*÷*M* for *D*, where the constant *M* represents the mass of the weight on the spring, and the constant *K* is a measure of the stiffness of the spring. The above identity is then written as follows:

```
DDPOS T                                           [6.3.26]
-(K÷M)×POS T
```

Consider the function

```
POS:(A×COS ω×(K÷M)*0.5)+B×SIN ω×(K÷M)*0.5
```

or equivalently

```
POS:+/(A,B)×2 1○ω×(K÷M)*0.5
```

for constants *A* and *B*. It is not difficult to show that identity 6.3.26 is valid for this function *POS*. Moreover, for *POS* and its derivative function *DPOS* we have:

```
POS 0                       DPOS 0
A                           B×(K÷M)*0.5
```

If the time *T*←0 represents the instant at which we begin describing the motion of the weight then evidently the scalar *A* represents the <u>initial</u> <u>position</u> of the weight and the scalar defined by *B*×(*K*÷*M*)*0.5 represents the <u>initial</u>

velocity of the weight. It is helpful to use the more mnemonic names *PI* for initial position and *VI* for initial velocity, and therefore we will substitute *PI* for *A* and *VI*÷(*K*÷*M*)⋆0.5 for *B* in our previous expression for the function *POS*. The function *POS* is then defined by the expression:

```
POS:+/(PI,VI÷(K÷M)*0.5)×2 1○ω×(K÷M)*0.5          [6.3.27]
```

Note that the function *POS* is a periodic function, whereas experience shows that the actual motion of the weight, while oscillatory, is not periodic because the distance it moves from the equilibrium position eventually becomes smaller and smaller. This damping effect, which is due to air resistance and other effects, has been neglected here. It is usual to assume that damping is proportional to velocity, and that damped vibrations are described by functions *POS* for which

```
D̲D̲POS T                                          [6.3.28]
-(C×D̲POS T)+(K÷M)×POS T
```

Solutions of this differential equation, which is similar in form to identity 6.2.13, can also be defined in terms of *E* ω, *SIN* ω, and *COS* ω. The analogies between electrical circuits and mechanical vibrations suggested by the similarity of identities 6.2.13 and 6.3.28 are an important part of the detailed analyses of these two applications. ⌹76

6.4 HYPERBOLIC FUNCTIONS

The primary functions denoted by 5○ω and 6○ω are defined by the identities

```
5○ω                          6○ω
0.5×(*ω)-*-ω                 0.5×(*ω)+*-ω
```

The function 5○ω is called the hyperbolic sine function (written sinh and pronounced cinch) and 6○ω is called the hyperbolic cosine function (cosh, for short). The functions 5○ω and 6○ω are called hyperbolic functions for reasons similar to those for calling 1○ω and 2○ω circular functions. Namely, just as all the points (1○*X*),2○*X* lie on a circle, all the points (5○*X*),6○*X* lie on a hyperbola.

Analysis of the hyperbolic functions will be carried out in the exercises. ⌹78

6.5 APPROXIMATE SOLUTIONS OF DIFFERENTIAL EQUATIONS

Each function so far introduced in this chapter is the solution of a differential equation, which itself can be used to directly produce the associated Taylor coefficient function. Taylor coefficient functions produced from differential equations are often too complex to be defined conveniently by a single expression. However, coefficient vectors of Taylor polynomials of specified order can be produced, and these Taylor polynomials provide approximations to the solutions.

To illustrate the use of Taylor polynomials in approximating solutions of differential equations we will use the motion of a vibrating bob on a spring (Section 6.3). For simplicity we will assume that both the mass of the bob and the stiffness constant of the spring are 1. Thus according to Hooke's law, the acceleration of the bob, as a function of position, is *-POS ω*. We will now include the damping effect due to air resistance by assuming that the acceleration diminishes by an amount proportional to the velocity. Acceleration is then a function of both position and velocity and is defined, for a positive scalar *C*, by:

ACC:-(*C*×*VEL ω*)+*POS ω* [6.5.1]

Even though definitions of the functions *POS* and *VEL* are unknown to us, we know that *DPOS*:*VEL ω* and *DVEL*:*ACC ω*, and therefore higher derivatives of *POS* and *VEL* can be produced formally from *DPOS* and *DVEL* as follows:

DDPOS:*DVEL ω* *DDDPOS*:*DDVEL ω*

and

DDVEL:*DACC ω* *DDDVEL*:*DDACC ω*

Also, since *ACC* is defined in terms of *POS* and *VEL* it can be formally differentiated as follows:

DACC:-(*C*×*DVEL ω*)+*DPOS ω* [6.5.2]

and

DDACC:-(*C*×*DDVEL ω*)+*DDPOS ω*

The approximation procedure is based on the fact that the values of all the above functions for an argument *T* can be produced from the values *POS T* and *VEL T*. For example, suppose that *POS* 5 is 1 and *VEL* 5 is ¯1 and *C* is 2. Then to produce the value *DDVEL* 5, we see that *DDVEL* 5 is *DACC* 5, which in turn is -(2×*DVEL* 5)+*DPOS* 5. *DVEL* 5 is *ACC* 5, which is -(2+*VEL* 5)+*POS* 5, or -(2×¯1)+1, or 1. *DPOS* 5 is *VEL* 5, or ¯1. Therefore *DDVEL* 5, which is -(2×*DVEL* 5)+*DPOS* 5, is -(2×1)+¯1, or ¯1. (It is instructive to continue this example and produce the values of all the above functions for the argument 5.)

Since we can produce values of all the above functions we can also produce values of the third order Taylor polynomial coefficient vector functions:

T3POS:(*POS ω*],(*DPOS ω*),(0.5×*DDPOS ω*),(÷6)×*DDDPOS ω*
T3VEL:(*VEL ω*),(*DVEL ω*),(0.5×*DDVEL ω*),(÷6)×*DDDVEL ω*

Continuing the above example, we have that *T3POS* 5 is 1 ¯1 0.5 ¯0.1667 and *T3VEL* 5 is ¯1 1 ¯0.5 0.1667.

Finally, since we can produce coefficient vectors of Taylor polynomials of *POS* and *VEL* at *T*, we can produce approximate values of *POS* and *VEL* for arguments near *T*. The functions

```
APOS:(T3POS α) POLY ω-α
AVEL:(T3VEL α) POLY ω-α
```

are the third order Taylor functions of *POS* and *VEL*.

Note that we could have defined derivatives of *POS* and *VEL* of orders higher than three and thus defined higher order Taylor functions of *POS* and *VEL*.

Let *P* be an increasing partition and assume for the moment that vectors of values *VPOS* and *VVEL* of *POS* and *VEL* respectively for the vector of arguments *P* are known. That is, *POS P*[*I*] is *VPOS*[*I*] and *VEL P*[*I*] is *VVEL*[*I*]. Then approximate values of *POS* and *VEL* for arguments in the interval with endpoints 1↑*P* and ¯1↑*P* can be produced from *APOS* and *AVEL*. Recall that the value *P IND X* of the function *IND* (Section 4.3), where *X* is in the interval with end points 1↑*P* and ¯1↑*P*, is the index of the subinterval of *P* containing *X*. For convenience we will assume that the vectors *VPOS* and *VVEL* are arranged in the two column matrix *TBL*, where *TBL*[;0] corresponds to *VPOS* and *TBL*[;1] corresponds to *VVEL*. The following are approximate, explicit definitions of *POS* and *VEL* on the interval with end points 1↑*P* and ¯1↑*P*:

```
POS:P[P IND ω] APOS ω : ω=P[P IND ω] : TBL[P IND ω;0]
VEL:P[P IND ω] AVEL ω : ω=P[P IND ω] : TBL[P IND ω;1]
```

Evidently *POS P*[*I*] is *VPOS*[*I*] while, in general, if *X* is in the *I*th subinterval of *P* then *POS X* is the value for *X* of the third order Taylor polynomial of *POS* centered at *P*[*I*]. Similarly for the function *VEL*.

Approximate vectors of values *VPOS* and *VVEL* can be produced from the Taylor polynomial functions *APOS* and *AVEL* if one value of *POS* and one value of *VEL* are known, the problem of producing the solution of a differential equation with one specified value is called an <u>initial value problem</u>. For example, suppose that *POS*0←1 is the value of *POS* at *P*[0] and *VEL*0←¯0.6 is the value of *VEL* at *P*[0], where *P*←0.5×⍳25. Also, *C*←0.75 where *C* is the constant in the definition of the acceleration function *ACC*. Define

```
BPOS:α[ω-1] APOS α[ω]
BVEL:α[ω-1] AVEL α[ω]
BROW:(α BPOS ω),α BVEL ω
BTB:ω,[0] α BROW 1↑ρω
```

Then approximate values of the elements of the table *TBL* are produced row-by-row as follows:

```
      TBL←1 2ρPOS0,VEL0

      3⍕TBL←P BTB TBL
1.000 ¯.600
 .729 ¯.488

      3⍕TBL←P BTB TBL
1.000 ¯.600
 .729 ¯.488
 .516 ¯.370
```

and so on. Thus if

```
      TABLE: α TABLE α BTB ω : (ρα)=1↑ρα BTABLE ω : α BTB ω
```

then approximate values of all elements of *TBL* are produced as follows:

```
      TBL←P TABLE 1 2ρ POS0,VEL0
```

Figure 6.5.2 is a graph of the approximate values *TBL*[;0] of the function *POS* for the vector of arguments *P*. ⊞81

As with other procedures, the question arises as to the accuracy of the approximations, and a rule of thumb similar to that for Newton's method is used. Specifically, suppose that *P*1 the vector of approximate values of *POS P* produced above, and that *P*2 is the vector of approximate values produced by the same procedure with *MP P* in place of *P* (*MP* is the midpoint function defined in Ex 4.41). Since ∧/*P*=(*MP P*)[2×ιρ*P*], both *P*1 and *P*2[2×ιρ*P*] are approximate values of *POS P*, and either of these vectors is taken as *K* decimal digit approximations to *POS P* if the elements of *P*1 and *P*2[2×ιρ*P*] agree to *K* digits.

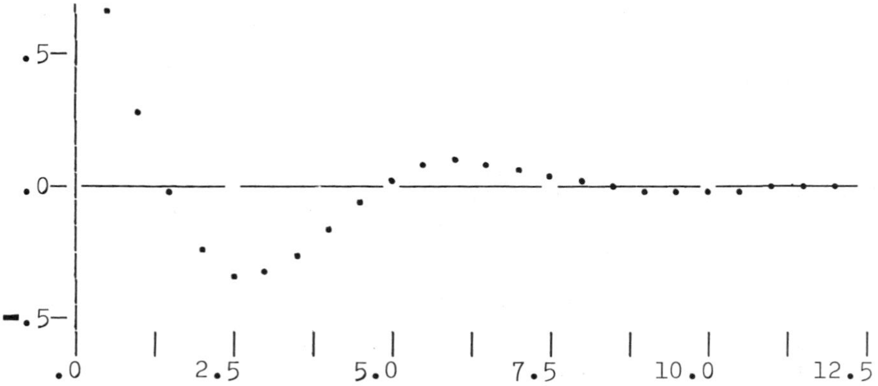

DAMPED HARMONIC MOTION
FIGURE 6.5.2

7

Inverse Functions

7.1 INTRODUCTION

The monadic functions *F* and *G* are said to be inverse functions of one another if there is a subset of the domain of *F* and a subset of the domain of *G* such that for all arguments *X* in the subset of the domain of *F* and all arguments *Y* in the subset of the domain of *G*:

G F X	*F G Y*	[7.1.1]
X	*Y*	

Also, *F* is said to be the inverse function of *G*, and *G* is said to be the inverse function of *F*, on the specified subsets of their domains. For example, the functions *F*:α⋆3 and *G*:ω⋆÷3 are inverses of one another on the set of all real numbers. The proof is as follows:

G F X	*F G Y*
(*X*⋆3)⋆÷3	(*Y*⋆÷3)÷3
X⋆3×÷3	*Y*⋆(÷3)×3
X	*Y*

It follows from the left side of 7.1.1 that if *Y* equals *F X*, then *X* equals *G Y*:

G Y	
G F X	Definition of *Y*
X	

Similarly, using the right side of 7.1.1 we see that if *X* equals *G Y*, then *Y* equals *F X*. In words, if *F* and *G* are inverse functions of one another then *Y* is the value of *F* for the argument *X* if and only if *X* is the value of *G* for the argument *Y*.

Although the concept of inverse functions is defined in terms of pairs of functions, in practice functions rarely appear in pairs of inverse functions. Usually one function, say *F*, is defined for some purpose and is then used to construct its inverse function $\underline{I}F$. There are several general procedures for constructing and analyzing the inverse function of a monadic scalar function *F*.

A graph of the inverse function $\underline{I}F$ of *F* can be produced directly from the graph of *F*. If the coordinate point *X*,*Y* lies on the graph of *F* then *Y* is the value of *F* for the argument *X*. Consequently *X* is the value of $\underline{I}F$ for the argument *Y*, which means that the coordinate point *Y*,*X* lies on the graph of *IF*. Therefore, the graph of *IF* is the reflection of the graph of *F* about the line with slope 1 passing through the origin. For example, the graph of the function α*3 is illustrated in Figure 7.1.2. The graph of the inverse function of α*3 can be produced as in Figure 7.1.3.

Tables of values of the inverse function *IF* can be constructed directly from tables of values of *F*. Suppose that *T* is a two column matrix for which *T*[;1] is the vector of values of *F* for the vector of arguments *T*[;0]. Therefore *T*[;0] is the vector of values of *IF* for the vector of arguments *T*[;1], which means that the matrix defined by ⌽*T* is a table of values of the inverse function of *F*. For example, the table

```
      T
 ¯2         ¯8
 ¯1         ¯1
 ¯0.5       ¯0.125
  0          0
  1          1
  1.5        3.375
  3         27
```

is a table of values for the function of α*3. That is, *T*[;1]=*T*[;0]*3. The matrix defined by ⌽*T* is a table of values of the function ω*÷3. That is, (⌽*T*)[;1]=(⌽*T*)[;0]*÷3.

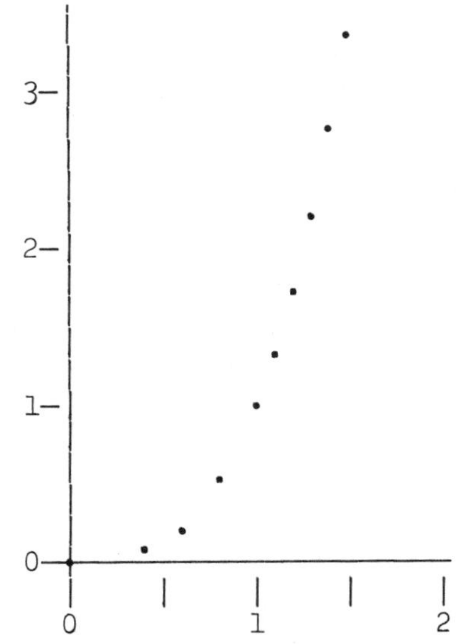

GRAPH OF α*3
FIGURE 7.1.2

Tables of an inverse function produced by the above procedure are often of little value because we have no control over the choice of the vectors of arguments of the inverse function. In order for a table-making procedure to be effective we must begin with vectors of arguments for an inverse function and then produce the corresponding vectors of values. If *Y* is a vector of arguments of the inverse function *IF* of *F*, then *Y* is a vector of values of *F*; in order to produce the corresponding vector of values *X* of *IF*, we must produce the vector of arguments *X* of *F* for which *Y=F X*. In other words, we must produce the solution *X* of the equation

Y=F X or equivalently of 0=*Y-F X*

Evidently element *I* of the vector *X* is a zero of the function *Y[I]-F* ω, which means that the vector *X* can be produced by a rootfinder. For example, using the discussion of root functions in Section 4.2 as a guide, the inverse function *IF* of the function *F*:α*3 can be defined for practical purposes as follows:

```
DF:3×α*2
ITTF:ω-(α-F ω)÷DF ω
RTF:α RTF α ITTF ω : 0.5E¯6>⌈/,|ω-α ITTF ω : α ITTF ω
IF:ω RTF ω
```

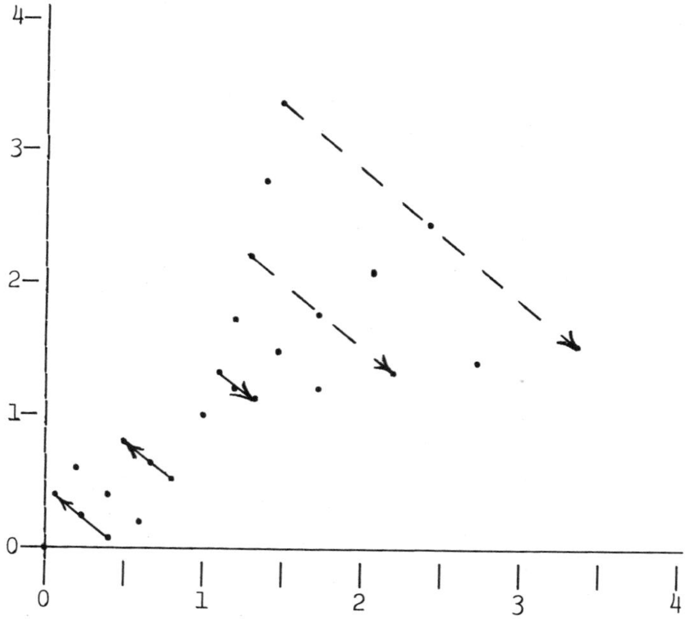

THE INVERSE FUNCTION OF α*3
FIGURE 7.1.3

The variations in these functions from those in Section 4.2 are due to the fact that we are seeking the root of the translate function *Y-F* α for each *Y*. Thus the tangent line intersection function *ITTF* and the root function *R̲TF* are dyadic functions (with left arguments representing *Y*), whereas they are monadic functions in Section 4.2. In Section 4.2 the argument of a root function is a starting value for Newton's method. Starting values are the right argument of *R̲TF*. Thus in the definition of *I̲F* starting values are chosen automatically; in the above example the starting value for *Y-F* α is *Y*. The geometry of most of the functions studied in this chapter is sufficiently simple to permit starting values to be chosen by simple functions of *Y*. For example:

```
      ⎕←X←I̲F ¯5 ¯3 2 7
¯1.709975947 ¯1.44224957 1.25992105 1.912931183

      X*3
¯5 ¯3 2 7                                                    ⌹1
```

If *F* and *G* are inverse functions with the derivatives *D̲F* and *D̲G* respectively, then according to the Composition Rule for Derivatives the derivative of the function *F G* ω is the function (*D̲F G* ω)×*D̲G* ω. Consequently the identity

```
      Y
      F G Y
```

implies the identity

```
      1
      (D̲F G Y)×D̲G Y
```

or equivalently

```
      D̲G Y                                          [7.1.4]
      ÷D̲F G Y
```

For example, if *F*:α*3 and *G*:ω*÷3 then *D̲F*:3×α*2 and consequently:

```
      ÷D̲F G Y
      ÷3×(Y*÷3)*2
      ÷3×Y*2÷3
      (÷3)×Y*-2÷3
```

According to 7.1.4, the last expression in this sequence of identities is a rule for the derivative function *D̲G* of *G*. That is, *D̲G*:(÷3)×ω*-2÷3.

The graphing techniques, the explicit definition of inverse functions in terms of a rootfinder, and Identity 7.1.4 provide a complete picture of inverse functions. There remains the question as to whether or not a particular monadic scalar function F has an inverse function.

The monadic function F is said to be <u>one-one</u> on a subset of its domain if for all arguments X and Y in the subset:

$$(X \neq Y) \leq (F\ X) \neq F\ Y$$

In words, if F is one-one then two distinct arguments cannot produce equal function values. If F is one-one on a subset of its domain, then F has an inverse function on that subset. The inverse function, which we name $\underline{I}F$, is defined formally by defining each value of $\underline{I}F$ as follows. The scalar Y is a member of the domain of $\underline{I}F$ if Y is a value of F for an argument X in that subset; the function value $\underline{I}F\ Y$ is obviously defined to be X. Since F is one-one, this definition of $\underline{I}F$ is not ambiguous. That is, there is one and only one possible choice for each function value $\underline{I}F\ Y$.

In particular, if F is a strictly monotonic function on the interval $(A \leq \omega) \wedge \omega \leq B$, then F has an inverse function on this interval.

The importance of inverse functions can be illustrated with simple differential equations. Suppose that H is a monadic scalar function and that the following identity is valid for the function F:

$$\underline{D}F\ \alpha$$
$$H\ F\ \alpha$$

Suppose also that one value of F is known, say $F\ A$ is B. If F has an inverse function G, then $G\ B$ is A, and substituting $G\ \omega$ for α in the above identity yields

$$\underline{D}F\ G\ \omega$$
$$H\ F\ G\ \omega$$

It then follows from Identities 7.1.1 and 7.1.4 that $\div\underline{D}G\ \omega$ equals $H\ \omega$, or equivalently $\underline{D}G\ \omega$ equals $\div H\ \omega$. Therefore G is an integral of $\div H\ \omega$. Thus if F is a known function, we can produce an integral of $\div H\ \omega$. Or, if an integral $C\ \omega$ of $\div H\ \omega$ is known, whose domain is an interval containing B, then G can be defined as $G{:}(A-C\ B)+C\ \omega$ on that interval, and thus we can produce the solution F of the above differential equation for which $F\ A$ is B. In either case, the inverse function G provides a link from a known function to a possibly new function.

7.2 THE INVERSE FUNCTION OF ⋆α

Since the function ⋆α is an increasing function on the interval of all real numbers, ⋆α has an inverse function on this interval. The range of values of ⋆α is the interval defined by ω>0; consequently the domain of the inverse function of ⋆α is the interval ω>0. The graph of the inverse function of ⋆α is illustrated in Figure 7.2.1.

We will denote the inverse function of ⋆α by *IE*. Consequently:

```
IE ⋆X              ⋆IE Y                                 [7.2.2]
X                  Y
```

The function *IE* can be defined for practical purposes as follows:

```
E:⋆ω
DE:⋆ω
ITTE:ω-(α-E ω)÷DE ω
RTE:α RTE α ITTE ω : 0.5E¯6>⌈/,ω-α ITTE ω : α ITTE ω
IE:ω RTE 1
```

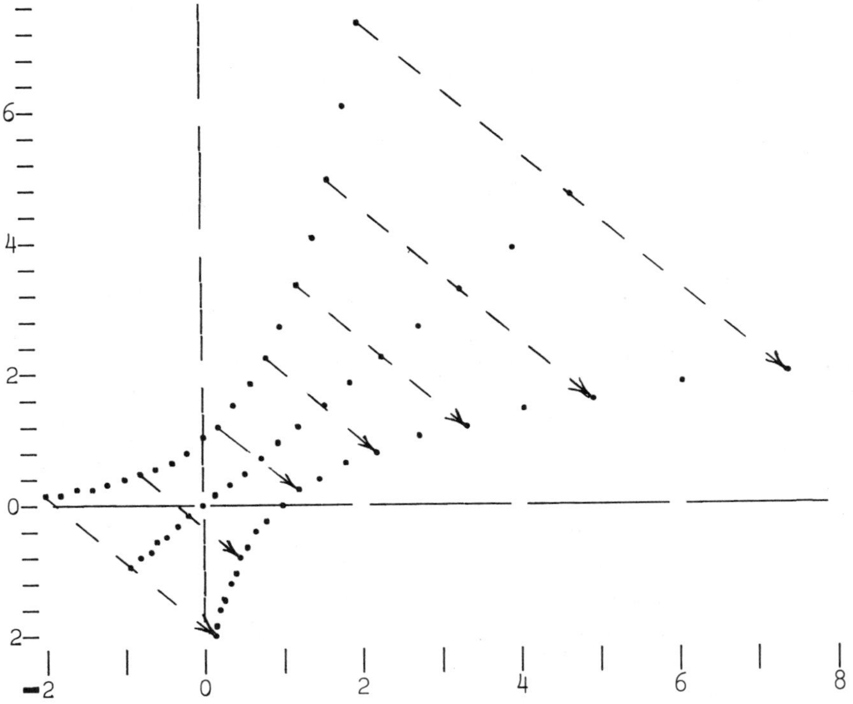

THE INVERSE FUNCTION OF ⋆α
FIGURE 7.2.1

For example:

```
      ⎕←X←IE .1 .5 2 13
¯2.302585093 ¯0.6931471806 0.6931471806 2.564949357

      ⋆X
0.1 0.5 2 13
```

For the function IE we have the logarithmic identity

```
IE ω×α                                              [7.2.3]
(IE α)+IE ω
```

The proof of Identity 7.2.3, which is based on 7.2.2 and the addition formula for ⋆α, will be discussed in the exercises. In view of 7.2.3, the function IE is called the (natural) logarithm function.

Identity 7.1.4 can be used to produce the derivative of the function IE. First, we substitute the function ⋆α for F and IE for G. Since ⋆α is identical to its derivative function, we also substitute ⋆α for the derivative DF. The result is the following identity for the derivative DIE:

```
DIE Y               since              ⋆IE Y
÷⋆IE Y                                 Y
```

these two identities lead to the identity

```
DIE Y                                               [7.2.4]
÷Y
```

In words, the derivative of the inverse growth function IE is the function ÷ω.

According to 7.2.4, the inverse growth function IE is an integral of ÷ω. Note that ÷ω is equivalent to ω⋆¯1 and that ω⋆¯1 is the only function of the form ω⋆N whose integral is not of the form C+(÷N+1)×ω⋆N+1 for a constant C. Since 1=⋆0 it follows that 0=IE 1. Thus, according to the First Fundamental Theorem of the Calculus,

```
IE ω
1 SLD ω
```

where SLD is the definite integral function of LD:÷ω (for logarithmic derivative). Thus if IE is defined for practical purposes as above, then SLD can be defined for practical purposes as

```
SLD:(IE ω)-IE α
```

Or, if SLD is defined for practical purposes as in Section 5.5, then IE can be defined for practical purposes as

```
IE:1 SLD ω
```

For example, using the latter definition of IE:

```
      ⋆IE 13
13
```

The Primary Function ⍟ω. The primary monadic scalar function denoted by ⍟ω and called the monadic logarithm function is defined by the identity

```
      ⍟ω
      IE ω
```

That is, the functions ⋆α and ⍟ω are inverses of one another. In terms of the function ⍟ω, Identities 7.2.2 and 7.2.3 are as follows:

```
      ⍟⋆X                 ⋆⍟Y                      [7.2.5]
      X                   Y
```
and
```
      ⍟ω×α                                         [7.2.6]
      (⍟α)+⍟ω
```

The latter identity is called a multiplication formula for the function ⍟ω. The following identities are proved from this multiplication formula.

```
      ⍟÷Y                                          [7.2.7]
      -⍟Y
```
and
```
      ⍟Y÷Z                                         [7.2.8]
      (⍟Y)-⍟Z                                           ⌹10
```

Polynomial Approximations to the Function ⍟ω. According to 7.2.4, the derivative of the function ⍟ω is the function ÷ω, or equivalently ω⋆¯1. The following table is a table of expressions for the higher derivatives of the logarithm function ⍟ω.

DERIVATIVE	EXPRESSION
0	⍟ω
1	ω⋆¯1
2	¯1×ω⋆¯2
3	2×ω⋆¯3
4	¯6×ω⋆¯4
I	(¯1⋆I-1)×(!I-1)×ω⋆-I

DERIVATIVES OF ⍟ω
TABLE 7.2.9

Consequently, if I>0 then element I of the coefficient vector of a Taylor polynomial of ⍟ω at the argument A is

```
      (¯1⋆I-1)×(!I-1)×(A⋆-I)÷!I
```

or equivalently

```
-((-A)⋆-I)÷I
```

Thus the family of coefficient vectors defined by the expression

```
(⍟A),-((-A)⋆-1+⍳α)÷1+⍳α
```

are the coefficient vectors of the Taylor polynomials of ⍟ω at the argument *A*, for which we have the sequence:

```
((⍟A),-((-A)⋆-1+⍳N)÷1+⍳N) POLY ω-A
(⍟A)+(0,-((-A)⋆-1+⍳N)÷1+⍳N) POLY ω-A
(⍟A)+(0,-(¯1⋆-1+⍳N)÷1+⍳N) POLY (ω-A)÷A
(⍟A)+(0,(¯1⋆⍳N)÷1+⍳N) POLY (ω-A)÷A
```

In summary, the Taylor coefficient function of ⍟ω is

```
TL:(⍟ω),-((-ω)⋆-1×⍳α+1)÷1+⍳α+1                    [7.2.10]
```

Equivalently,

```
(⍟A)+(0,(¯1⋆⍳N+1)÷1+⍳N+1) POLY (ω-A)÷A           [7.2.11]
```

is the *N*th order Taylor polynomial of ⍟ω at *A*. The latter expression is used in the exercises to analyze the Taylor polynomials of ⍟ω. ⍠1

<u>The Dyadic Power and Logarithm Functions</u>. For a constant *R* such that 0≠*R*, the function ⋆*R*×α is a strictly monotonic function on the interval of all scalars, and therefore has an inverse function on this interval. The inverse function of ⋆*R*×α is the function (⍟ω)÷*R*. That is:

```
(⍟⋆R×X)÷R            ⋆R×(⍟Y)÷R                     [7.2.12]
X                    Y
```

The proofs are as follows:

```
(⍟⋆R×X)÷R            ⋆R×(⍟Y)÷R
(R×X)÷R              ⋆⍟Y
X                    Y
```

The primary <u>dyadic power function</u> denoted by α⋆ω is defined by the identity

```
α⋆ω                                                [7.2.13]
⋆(⍟α)×ω
```

and the primary <u>dyadic logarithm function</u> denoted by α⍟ω is defined by the identity

```
α⍟ω                                                [7.2.14]
(⍟ω)÷⍟α
```

For a scalar *B* for which *B*>0, the function *B*⋆α and the function *B*⍟ω are inverse functions. The proof of this fact will be discussed in the exercises.

It must be shown that the above definition of the dyadic power function produces the expected function. Namely, we must show that

```
A⋆B                                          [7.2.15]
×/BρA
```

for non-negative integers *B*, and that

```
A⋆B+C                                        [7.2.16]
(A⋆B)×A⋆C
```

The addition formula 7.2.16 follows directly from the addition formula for ⋆α. The proof of Identity 7.2.15 is:

```
A⋆B
⋆(⍟A)×B                      Identity 7.2.13
⋆+/Bρ⍟A                      Assumption on B
×/Bρ⋆⍟A                      Add. Form. for ⋆α
×/BρA
```

⌹22

7.3 INVERSE FUNCTIONS OF α⋆*N*

For a positive integer *N*, the function α⋆*N* is increasing:

On the interval of all real numbers if *N* is odd;
On the interval ω≥0 if *N* is even.

The function ω⋆÷*N* is the inverse function of α⋆*N* on this interval. Graphs and tables of values for these inverse functions will be discussed in the exercises. ⌹25

The derivative of the function ω⋆÷*N* can be determined from 7.1.4. Substituting α⋆*N* for *F* and *N*×α⋆*N*-1 for $\underline{D}F$ and ω⋆÷*N* for *G* in the second expression of that identity yields

```
÷N×(Y⋆÷N)⋆N-1
```

for which we have the following sequence:

```
÷N×(Y⋆÷N)⋆N-1
÷N×Y⋆(N-1)÷N
÷N×Y⋆1-÷N
(÷N)×÷Y⋆1-÷N
(÷N)×Y⋆(÷N)-1
```

Therefore, the derivative of the function α⋆÷*N* is the function (÷*N*)×α⋆(÷*N*)-1.

We can now determine the derivative of the function α⋆M÷N, where M and N are integers such that (0≠M)∧0<N. Suppose that F:ω⋆÷N and G:ω⋆M. Then D̲F:(÷N)×ω⋆(÷N)-1 and D̲G:M×ω⋆M-1. Since (X⋆M)⋆÷N equals X⋆M÷N, we also have the identity

```
X⋆M÷N
F G X
```

According to the Composition Rule for Derivatives, the derivative of the function F G ω is the function (D̲F G ω)×D̲G ω, for which we have the following sequence:

```
(D̲F G ω)×D̲G ω
((÷N)×(ω⋆M)⋆(÷N)-1)×M×ω⋆M-1
(M÷N)×(ω⋆(M÷N)-M)×ω⋆M-1
(M÷N)×ω⋆(M÷N)+(-M)+M-1
(M÷N)×ω⋆(M÷N)-1
```

Therefore the derivative of the function ω⋆M÷N is the function (M÷N)×ω⋆(M÷N)-1.

We have just shown that for a non-zero rational number R the derivative of the function ω⋆R is the function R×ω⋆R-1. In fact, this is true for any non-zero scalar R. To see that this is true, recall that ω⋆R is equivalent to ⋆R×⍟ω (Identity 7.2.13), and the derivative of the latter expression is (⋆R×⍟ω)×R÷ω, for which we have the following sequence of identities:

```
(×R×⍟ω)×R÷ω
(ω⋆R)×R÷ω
R×(ω⋆R)×ω⋆¯1
R×ω⋆R-1
```

7.4 INVERSE CIRCULAR FUNCTIONS

The sine function denoted by 1○ω is an increasing function on the interval (○0.5)≥|ω. Recall that the functions 1○ω and 2○ω are defined in such a way that the vector 2 1○X represents the point on the unit circle which is the arclength X away from the point 1 0. For an argument Y for which 1≥|Y the value of the inverse function of the sine function is the arclength or arc X for which Y=1○X. Consequently the inverse function of the sine function is called the arc of the sine function and is usually named *ARCSIN*.

Graphs and tables of values of the function *ARCSIN* will be discussed in the exercises. The derivative of this function can be obtained from 7.1.4. Since the derivative of the function 1○ω is the function 2○ω, that identity yields:

```
D̲ARCSIN Y
÷2○ARCSIN Y
```

Since the values of the function *ARCSIN* lie in the interval (○0.5)≥|ω, the values of the function 2○*ARCSIN* ω are non-negative. Consequently:

```
2○ARCSIN Y
(1-(1○ARCSIN Y)*2)*0.5
```

Combining these two identities yields the following sequence of identities for the derivative function *D̲ARCSIN*:

```
D̲ARCSIN Y
÷2○ARCSIN Y
÷(1-(1○ARCSIN Y)*2)*0.5
÷(1-Y*2)*0.5
÷0○Y
```

That is:

```
D̲ARCSIN Y                                [7.4.1]
÷0○Y
```

This identity will be applied to the analysis of the function *ARCSIN* in the exercises.

The sine function 1○ω is strictly monotonic on each interval (ω≥○*M*-0.5)∧ω≤○*M*+0.5, where *M* is an integer, and therefore has an inverse function on each such interval. A careful examination of the graph of the sine function will show that the inverse function of the sine function on the interval (ω≥○*M*-0.5)∧ω≤○*M*+0.5 can be defined by the expression

```
(○M)+ARCSIN ω×¯1*M
```

The primary monadic scalar function denoted by ¯1○ω is defined by the identity

```
¯1○ω
ARCSIN ω
```
⌹30

The inverses *ARCCOS* of *COS* and *ARCTAN* of *TAN* will be discussed in the exercises. The pattern of denoting inverse primary circular functions suggested by 1○α and ¯1○ω extends, so that ¯2○ω is the inverse of 2○α and is therefore *ARCCOS* ω, and so on. ⌹33

7.5 INVERSE HYPERBOLIC FUNCTIONS

The inverse hyperbolic functions can be constructed by the procedure described in Section 7.1. It is also possible to construct explicit rules for these functions in terms of the monadic logarithm function.

For example, the hyperbolic sine function denoted by 5○α is an increasing function of the interval of all scalars and therefore has an inverse function on this interval. The inverse function of the hyperbolic sine function is called *ARSINH*.

To construct an explicit rule for the function *ARSINH* we must construct an expression whose value for an argument *Y* is identical to the solution *X* of the equation *Y*=5○*X*, or equivalently:

```
Y=0.5×(*X)-*-X                                        [7.5.1]
```

It is helpful to let *U*←**X*. Then (*-*X*)=÷*U*. Substituting *U* for **X* and ÷*U* for *-*X* in Equation 7.5.1 yields the equation

```
Y=0.5×U-÷U                                            [7.5.2]
```

Then if *U* is a positive solution of Equation 7.5.2, ⍟*U* is a solution of Equation 7.5.1. The following sequence of equations are equivalent to Equation 7.5.2:

```
Y=0.5×U-÷U
(2×Y)=U-÷U
(2×Y×U)=(U*2)-1
0=¯1+(¯2×Y×U)+U*2
```

For each *Y* the function ¯1+(¯2×*Y*×ω)+ω*2 is a polynomial of degree 2. The roots of this polynomial are the solutions of the last equation in the above sequence, and thereby the solutions of 7.5.2. The roots of this polynomial are 0.5×(-¯2×*Y*)+(4+4×*Y**2)*0.5 and 0.5×(-¯2×*Y*)-(4+4×*Y**2)*0.5, or equivalently, *Y*+4○*Y* and *Y*-4○*Y*. It is not difficult to check that the values of the expression *Y*+4○*Y* are positive for all arguments *Y*, while the values of *Y*-4○*Y* are negative; consequently ⍟*Y*+4○*Y* is the solution of Equation 7.5.1. Therefore:

```
ARSINH Y
⍟Y+4○Y
```

The primary monadic scalar function denoted by ¯5○ω is defined by the identity

```
¯5○ω                                                  [7.5.3]
ARSINH ω
```

Consequently:

```
      ¯5○⍵
      ⍟⍵+4○⍵
```

For example:

```
      Y
¯1 1.5 3 7

      ¯5○Y
¯0.881373587 1.194763217 1.818446459 2.644120761

      ⍟Y+4○Y
¯0.881373587 1.194763217 1.818446459 1.644120761
```
⌹42

The inverse hyperbolic functions called *ARCOSH* and *ARTANH* will be discussed in the exercises. ⌹44

7.6 INVERSE LINEAR FUNCTIONS

The matrices *M* and *N* are said to be <u>inverses of one another</u> if

```
      M+.×N                    N+.×M
      ID 1↑⍴M                  ID 1↑⍴N
```

Also, *N* is said to be the <u>inverse matrix</u> of *M* and *M* is said to be the <u>inverse matrix</u> of *N*.

A matrix *M* has an inverse matrix if and only if the function *M*+.×⍵ has an inverse function; *M*+.×⍵ has an inverse function if and only if *M*+.×⍵ is one-one. Thus *M* has an inverse matrix if and only if for each conformable vector *Y*, the equation ∧/*Y*=*M*+.×*X* has one and only one solution *X*. Consequently, a square matrix *M* has an inverse matrix if and only if *M* is a valid right argument of the dyadic function called domino. The inverse matrix of *M* can be produced by the monadic domino function. That is, if a square matrix *M* has an inverse, then the inverse is ⌹*M*.

```
      M+.×⌹M                   (⌹M)+.×M                [7.6.1]
      ID 1↑⍴M                  ID 1↑⍴M
```

For example:

```
      M                              1⍕⌹M
1 1 1                          3.0 ¯3.0  1.0
1 2 4                         ¯2.5  4.0 ¯1.5
1 3 9                           .5 ¯1.0   .5

      0⍕M+.×⌹M                       0⍕(⌹M).×M
1  0  0                        1  0  0
0  1  0                        0  1  0
0  0  1                        0  0  1
```

If the square matrix *M* has an inverse, then the function M+.×⍵ has an inverse function, which is (⌹M)+.×⍺. To prove this fact we must show that the identities

```
(⌹M)+.×M+.×X                M+.×(⌹M)+.×Y           [7.6.2]
X                           Y
```

are valid. The proof of the left side is as follows:

```
(⌹M)+.×M+.×X
((⌹M)+.×M)+.×X              Associativity of +.×
(ID 1↑⍴M)+.×X               Identity 7.6.1
X                           Definition of ID
```

The proof of the right side is similar and is left to the reader. For each *Y* the value (⌹M)+.×Y is identical to the solution of the equation

```
∧/Y=M+.×⍵
```

The solution of this equation can also be produced by the primary dyadic domino function, and is Y⌹M. Consequently we have the following identity relating the monadic domino function and the dyadic domino function:

```
(⌹M)+.×Y                                           [7.6.3]
Y⌹M
```

If *F* is a linear function which applies to vectors then *F* can be represented as a +.× inner product as follows:

```
F X
(E̲F ID ⍴X)+.×X
```

If the matrices appearing in this representation are square matrices that have inverse matrices, then the function *F* has an inverse function I̲F which also applies to vectors and that can be defined as follows:

```
I̲F:⍵⌹E̲F ID ⍴⍵
```

For example, the function F:(-\⍺) is a linear function which applies to vectors and for which the matrices appearing in its +.× representation are complete lower triangular matrices. These matrices are valid right arguments of domino, and therefore have inverse matrices. Consequently, this function has an inverse function I̲F:⍵⌹E̲F ID ⍴⍵.

In fact, the function -\⍵ applies along the first axis of its arguments and therefore we need not use the extended domain function to define its inverse function. That is, the inverse function of -\⍵ can be defined as ⍵⌹-\ID ⍴⍵. ⌹5

8 Differentiation and Integration

8.1 INTRODUCTION

The rules of differentiation and several rules of integration were introduced in Chapter 1. We will now formalize the application of the derivative rules and introduce new integral rules.

The rules of differentiation have so far been used in an informal way, in that no formal procedure has been established for selecting the appropriate rule to be applied in each case. A formal selection procedure is developed in the next section, and is used in the definition of a function that produces derivatives.

Integration is not as simple or direct a procedure as differentiation. For instance, there is more than one integral rule that might apply to the product of two functions. It is therefore more difficult to formalize the application of integral rules. Also, derivatives of functions defined in terms of the elementary functions (1○ω, *ω, ω*2, etc.) can always be expressed in terms of elementary functions, but this is not true of integrals. As a result, the procedures for producing integrals do not always work. You try one, and if that fails you try another. Because of this difficulty extensive reference tables of integrals have been developed, and there are many sources of such tables. Thus, while it is useful to become familiar with the basic procedures for producing integrals (Sections 8.3-6), there is little need to become an expert.

In order to formalize the application of the derivative rules we must be able to analyze function definitions. Expressions defining functions can be thought of as strings of characters, as in 3+ω*4, where the characters are 3,+,ω,*, and 4. Note that the characters 3 and 4 are distinguished from the numbers represented by 3 and 4, and the characters + and * are distinguished from the functions whose names are + and *. These distinctions are formalized by enclosing characters in quotes. Thus '3' is the character 3, as opposed to the number represented by 3, and '+' is the character +, as opposed to the name of the plus function. Also, '3+ω*4' is a string of characters, as opposed to the function definition 3+ω*4. The string of characters '3+ω*4' is called a <u>representation</u> of the function defined by 3+ω*4.

Selection functions apply to strings of characters, and we speak of <u>vectors of characters</u> or <u>character vectors</u>, character matrices, and so on. For example:

```
      A←'3+ω*4'
      A
3+ω*4
      ⌽A
4*ω+3
      ρA
5
      2 5ρA
3+ω*4
3+ω*4
      1 0 1 0 1/A
3ω4
```

Values of expressions represented by character vectors are produced by the execute function ⍎. For example:

```
      ⍎'3+4'
7
      ⍎'DIFF 3 7 19'
4 12
```

The representation of an array of numbers as a character array is produced by the format function ⍕. For example:

```
      DIFF 3 7 19
4 12
      ρDIFF 3 7 19
2
      ⌽DIFF 3 7 19
12 4
```

```
      ⍕DIFF 3 7 19
4 12
      ρ⍕DIFF 3 7 19
4
      ⌽⍕DIFF 3 7 19
21 4
```

There are two primary functions not yet introduced which will be useful in formalizing the application of the derivative rules. The dyadic function α∊ω is called membership; its value *A*∊*B* is the same size as *A*, with element 1 if the corresponding element of *A* is also on element of *B*, and 0 otherwise. For example:

```
      3 4 ¯5 0 ∊ 0 4
0 1 0 1
      'ABCDEFG'∊'DAF'
1 0 0 1 0 1 0
```

The dyadic function α⍳ω is called index. The left argument of index is a vector. The value *V*⍳*A* is the same size as *A*, with elements equal to the indices of the first occurrences of the elements of *A* in *V*, and ρ*V* otherwise. For example:

```
      0 4⍳3 4 ¯5 0
2 1 2 0
      ρ0 4
2
```

```
      'DAF'⍳'ABCDEFG'
1 3 3 0 3 2 3
      ρ'DAF'
3
```

⌹1

8.2 A DERIVATIVE FUNCTION

We will now define a monadic function D̲ such that:

> The arguments of D̲ are to be vectors of characters representing functions;
>
> The values of D̲ will also be character vectors representing functions. The function represented by a value of D̲ is to be equivalent to the derivative of the function represented by the corresponding argument of D̲.

The function D̲ is defined recursively in terms of the derivative rules. For example, consider the monadic scalar function represented by the character vector '(1○α)+2○α'. The derivative of this function can be produced in terms of the dyadic plus rule. Evidently, the derivative is identical to the derivative of the function 1○α plus the derivative of 2○α. Consequently it must be true that the function represented by the character vector

```
D̲ '(1○α)+2○α'
```

is equivalent to the function represented by the character vector defined as follows:

```
'(',(D̲ '1○α'),')+',D̲ '2○α'
```

If we are to use the derivative rules in the definition of the function D̲, we must first define a function which determines the appropriate rule to be applied in each case. The selection of a derivative rule is based on the <u>factorization</u> of the function whose derivative is being produced. That is, in selecting a derivative rule to be applied to a function *H* we first determine a primary scalar function f and either one monadic scalar function *G* or two monadic scalar functions *F* and *G* such that either:

```
H ω        or        H ω
fG ω                 (F ω)fG ω
```

In the first case we apply the monadic f derivative rule and in the second case the dyadic f derivative rule.

In either form of the factorization of *H* described by the above identities, *H* is equivalent to an expression such that whenever this expression is evaluated, the primary function f is evaluated last. Thus to determine an appropriate derivative rule for a character vector that is an argument of D̲, we can determine the symbol (and its location) of a primary scalar function that could be evaluated last whenever the function represented by this character string is evaluated.

In any expression without redundant outside parentheses, as in (*A*+*B*), the only candidates for the last evaluated function are those which are not imbedded within parentheses. For example, when evaluating the expression

```
((A+B)*C)-A×(B÷C)
```

the function ÷ is evaluated before × and + is evaluated before *, which is in turn evaluated before -. Finally, × is evaluated before -, so that - is the last function evaluated. The function

```
LP←'('
RP←')'
DEPTH:+\(ω=LP)-0,¯1↓ω=RP
```

can be used to determine those primary functions in an expression which are not imbedded in parentheses. *DEPTH* is a monadic function whose arguments are character vectors and whose values are vectors of integers indicating the depth of imbedding within parentheses of each character in the corresponding argument. For example:

```
      DEPTH '((A+B)*C)-A×(B÷C)'
1 2 2 2 2 2 1 1 1 0 0 0 1 1 1 1 1
```

The characters at depth 0 can be produced as follows:

```
      S←'((A+B)*C)-A×(B÷C)'

      (0=DEPTH S)/S
-A×
```

The primary scalar functions at depth 0 can be produced as follows:

```
      ((0=DEPTH S)∧S∊'+-×÷*')/S
-×
```

Among the primary functions in an expression, the one that will be evaluated last is the one at depth 0 which is furthest to the left. The index of this function in the character vector representing the expression can be produced as follows:

```
      ⎕←I←((0=DEPTH S)∧S∊'+-×÷*')⍳1
9
      S[I]
-
```

The primary scalar functions with which we will be concerned for the derivative algorithm are +,-,×,÷,*,⍟, and ○. The index of the last evaluated function from this list in an expression is produced by the monadic function

```
PFS←'+-×÷*⍟○'
PF:((0=DEPTH ω)∧ω∊PFS)⍳1
```

For example:

```
      PF S
9
```

and

```
      ⎕←I←PF '(1○α)×2○α'
5
      '(1○α)×2○α'[I]
×
```

For a character vector *S* that represents the function named *H*, define the integer *I* and the character vectors *U* and *V* as follows:

```
I←PF S
U←I↑S
V←(I+1)↓S
```

For example:

```
      I←PF S←'(1○α)×2○α'

      ⎕←U←I↑S
(1○α)
      ⎕←V←(I+1)↓S
2○α
```

The character *S*[*I*] is the symbol of a primary scalar function f and *V* represents a function which we will call *G*. If the vector *U* is empty then

```
H ω
fG ω
```

and to produce the derivative of *H ω* we apply the monadic f derivative rule to the expression f*G ω*. If *U* is not empty then *U* represents a function which we will call *F* and

```
H ω
(F ω)fG ω
```

In this case we produce the derivative of *H ω* by applying the dyadic f derivative rule to the expression (*F ω*)f*G ω*.

Up to now we have been concerned only with expressions containing certain primary functions. However, derivatives of expressions containing names of secondary functions can also be produced, and our derivative function should also apply to them. In order to apply the derivative function to these expressions we must be able to locate the function, primary or secondary, evaluated last in an expression. There is a simple criterion for locating this function when all secondary functions in the expression are monadic. Namely, we first locate the leftmost name or constant at depth 0. If this is the name of a secondary function, and if it appears to the left of the primary function located by *PF*, then it is the name of the function evaluated last. Otherwise, the primary function located by *PF* is the one evaluated last. The function

```
      BL←' '
      AR←'αω'
      NL:(0=DEPTH ω)∧~ω∊PFS,BL,AR
```

locates function and variable names and constants at depth 0 in an expression. For example:

```
      S←'FG(ω+ATV)-A*D3'
      I←NL S
      I\I/S
FG          A D3
```

The function

```
      SFN:((NL ω)∧1≥0,+\|DIFF NL ω)/ιρω
```

locates the leftmost name or constant at depth 0. Continuing the above example:

```
      SFN S                          S[SFN S]
0 1                            FG
```

Whether or not a name is the name of a secondary function can be determined by the function denoted by ⎕*NC*; that is, *G* is the name of a secondary function if ⎕*NC* '*G*' is 3. Also, if *G* is a variable name then ⎕*NC* '*G*' is 2. For example:

```
      ⎕NC 'PF'                       ⎕NC 'BL'
3                              2
```

The function

```
      SF:SFN ω : 3≠⎕NC ω[SFN ω] : ρω
```

tests whether the leftmost name or constant at depth 0 is the name of a secondary function; if it is, the result of *SF* is the vector of indices to the name, and otherwise the result is the length of the expression. For example, none of the names in the above vector *S* are names of secondary functions:

```
      SF S                           ρS
14                             14
```

On the other hand, the vector `'DIFF I/ω'` does contain the name of a secondary function:

```
      SF 'DIFF I/ω'
0 1 2 3
      'DIFF I/'[0 1 2 3]
DIFF
```

Finally, we have

$$LAST:SF \ \omega \ : \ \wedge/(PF \ \omega) \le SF \ \omega \ : \ PF \ \omega$$

If the character vector argument of *LAST* does not contain names of secondary, dyadic functions then the result of *LAST* is the location of the primary or secondary function evaluated last.   For example, consider the expression $AB-\omega*2$. If *AB* is a variable name, then

```
      LAST 'AB-ω*2'
2
      'AB-ω*2'[2]
-
```

However, suppose that *AB* is a function name; for instance, $AB:2+\omega$. Then

```
      LAST 'AB-ω*2'
0 1
      'AB-ω*2'[0 1]
AB
```

Note that before applying the function *LAST*, we must remove excess outside parentheses from its argument, for otherwise no element of the argument is at depth 0. This is done by the following functions:

```
      RBLA:(~∧\B̲L̲=ω)/ω
      RBL:ΦRBLA ΦRBLA ω
      STRIP:RBL ω : ∧/0≠DEPTH RBL ω : STRIP 1↓¯1↓RBL ω
```

For example:

```
      STRIP ' (  (A+B )  )'
A+B
```

The function *LAST* is the basis of the definition of the function *RULE*, which selects the appropriate derivative rule to be applied in each case. The derivative rules are themselves functions, whose names will be D̲0, D̲1, through D̲14. The order of these names reflects the order of primary function symbols in the vector P̲F̲S̲. Specifically, D̲0 is the monadic plus rule, D̲1 the dyadic plus rule, D̲2 the monadic minus rule, and so on. D̲14 is to be applied whenever *LAST* locates the name of a secondary function.

The functions coresponding to monadic derivative rules are monadic functions, and those corresponding to dyadic rules are dyadic functions. For example, *D* is applied to '-ω*2' by evaluating *D*2 'ω*2' (the monadic minus rule), and *D* is applied to '3+ω*5' by evaluating '3' *D*1 'ω*5' (the dyadic plus rule).

The left and right arguments of the derivative rule functions are produced by the functions *LA* and *RA* respectively:

```
QT←''''
```

(the value of *QT* is the single quote character itself, that is, ')

```
ENQ:QT,ω,QT
LA:(0≠1↑ω)/ENQ (1↑ω)↑α
RA:ENQ (1+¯1↑ω)↓α
```

For example:

```
      '3+ω*5' LA LAST '3+ω*5'
'3'
      '3+ω*5' RA LAST '3+ω*5'
'ω*5'
```

The name of the appropriate derivative rule function is produced by the function

```
DS←'D'
RLN:DS,⍕(2×PFSι1↑α[ω])+0≠1↑ω
```

For example:

```
      '3+ω*5' RLN LAST '3+ω*5'
D1
```

which is the name of the dyadic plus rule. Finally, we have

```
RULE:(ω LA LAST ω),BL,(ω RLN LAST ω),BL, ω RA LAST ω
```

For example:

```
      RULE '3+ω*5'
'3' D1 'ω*5'
```

Thus *D* must execute the results of *RULE*.

Since *D* is to be defined recursively, we must also have derivative rules at which the recursion stops. That is, we must have rules whose definitions do not depend on other values of *D*. One such rule is that whenever the argument *S* of *D* is of length 1 and contains either α or ω, then the value *D S* is '1'; this corresponds to the fact that

if *F*:α or *F*:ω, then *D̲F*:1+0×α. The other rule w
depend on other values of *D̲* is if *S* contains ne
ω, then the value *D̲ S* is '0'. The function

```
CONST:(0≠ρω)∧∧/~AR∊ω
```

produces the value 1 if its argument is not empty a
not contain either character 'α' or 'ω', and 0 othe
The function

```
IDENT:(1=ρω)∧∨/AR∊ω
```

produces the value 1 if its argument is either 1ρ'α' o
1ρ'ω', and 0 otherwise. These two functions are combined into one by the function

```
END:(CONST ω),IDENT ω
```

For example:

```
      END '3+*45'
1 0
      END 'α'
0 1
      END 'A×α÷3'
0 0
```

Thus if *END S* is not 0 0, the value *D̲ S* is (*END S*)/'01'. The derivative function *D̲* is defined as follows:

```
MA←'+0×ω'
MD:ω,(CONST ω)/MA

DE←'01'
D:MD ⍎RULE STRIP ω : ∨/END STRIP ω : (END STRIP ω)/DE
```

(The use of *MD̲* in the definition of *D̲* ensures that the result will be the definition of a monadic function.)

Before we can apply the derivative function *D̲* we must define the derivative rule functions *D̲*0 through *D̲*14. For that purpose we define

```
ENP:LP,ω,RP
```

which encloses its argument in parentheses. Then the dyadic plus rule is

```
D1:(ENP D α),PFS[0],D ω
```

For example:

```
      D '3+ω'
(0)+1+0×ω
```

ιso, the monadic and dyadic minus rules are

```
D2:PFS[1],D ω
D3:(ENP D α),PFS[1],D ω
```

The remaining primary function derivative rules will be developed in the exercises.

We turn now to the definition of the chain rule function *D*14. We might try the following definition:

```
D14:(ENP DS,α,ω),PFS[2],D ω
```

For example, recalling that *AB* is a defined function:

```
      D 'AB -ω'
(DAB -ω)×-1
```

Suppose that we want to apply *D* to the above result. In the process *D* will be applied to the expression *DAB -ω*:

```
      D 'DAB -ω'
(0)+-1+0×α
```

Evidently this is not the expected result. The problem is that, even though we consider *DAB* to be a function name, no function with that name has been defined. Consequently *DAB* is considered a variable name in the function *D*. This difficulty is overcome by defining *D*14 in such a way that it also fixes the definitions of derivatives which it produces.

If the character vector *S* represents the αω form of a function definition, then *DERA S* represents the αω form of its derivative, where:

```
CL←':'
DERA:DS,((1+ωιCL↑ω),D (1+ωιCL)↓ω
```

For example:

```
      DERA 'F:2+ω'
DF:(0)+1+0×α
```

A function definition in the αω form is fixed (and its name is displayed) by the function *BDEF*9 from Appendix A. Therefore we define

```
DERB:BDEF9 DERA ω
```

For example:

```
      DERB 'F:2+ω'
DF
      DF 1 2 3
1 1 1
```

If the argument of the function *CDEF*9 from Appendix A represents the name of a defined function, then the result is the αω form of its definition. For example, if *G*:ω-7 then

```
      CDEF9 'G'
G:ω-7
```

Using *CDEF*9, we define

```
      DER:DERB CDEF9 ω
```

Continuing the above example:

```
      DER 'G'
D̲G
      D̲G ι5
1 1 1 1 1
```

Thus *D̲*14 is defined as follows:

```
      D̲14:(ENP (DER α),ω),P̲F̲S̲[2],D̲ ω
```

Repeating the example at the beginning of the paragraph:

```
      D̲ 'AB-ω'
(D̲AB -ω)×-1

      D̲ 'D̲AB -ω'
(D̲D̲AB -ω)×-1
```

We conclude this section by describing the changes to be made in the functions in Appendix A so that derivatives can be produced within the function *DEF*. The effect of these changes will be, if *F* is a defined function, that entering *D̲F* within the function *DEF* will result in the display of the αω form of the definition of *D̲F*, whether or not *D̲F* has been previously defined. Moreover, if *D̲F* has not been previously defined, then its definition will be fixed.

At present, the result of entering *D̲F* within the function *DEF* is the αω form of the definition of *D̲F* only if *D̲F* has been previously defined, and is the empty vector otherwise. These results are produced by the function *CDEF*9, which will be redefined as follows, so as to produce the value of a new function *ODEF*9 whenever its argument does not represent the name of a defined function:

```
      CDEF9:⍎R9,STL9 ω : 3≠⎕NC ω : ODEF9 STL9 ω
```

The function *ODEF*9 produces derivatives under the appropriate conditions:

```
      ODEF9:CDEF9 DER 1↓ω : (D̲S̲≠1↑ω)∨3≠⎕NC 1↓ω : 0ρω
```

For example:

```
      DEF
H:ω+1

      DEF
DH
DH:(1)+0+0×ω
      DH 7 10
1 1
```
⌹7

The following is a summary of the functions and variables defined in this section:

```
      LP←'('
      RP←')'
      DEPTH:+\(ω=LP)-0,¯1↓ω=RP
      PFS←'+-×÷*⍟○'
      PF:((0=DEPTH ω)∧ω∊PFS)⍳1
      BL←' '
      AR←'αω'
      NL:(0=DEPTH ω)∧~ω∊PFS,BL,AR
      SFN:((NL ω)∧1≥0,+\|DIFF NL ω)/⍳ρω
      SF:SFN ω:3≠⎕NC ω[SFN ω]:ρω
      LAST:SF ω :  ∧/(PF ω)≤SF ω : PF ω
      RBLA:(~∧\BL=ω)/ω
      RBL:⌽RBLA ⌽RBLA ω
      STRIP:RBL ω : ∧/0≠DEPTH RBL ω : STRIP 1↓¯1↓RBL ω
      QT←''''
      ENQ:QT,ω,QT
      LA:(0≠1↑ω)/ENQ (1↑ω)↑α
      RA:ENQ (1+¯1↑ω)↓α
      DS←'D'
      RLN:DS,⍕(2×PFS⍳1↑α[ω])+0≠1↑ω
      RULE:(ω LA LAST ω),BL,(ω RLN LAST ω),BL,ω RA LAST ω
      CONST:(0≠ρω)∧∧/~AR∊ω
      IDENT:(1=ρω)∧∨/AR∊ω
      END:(CONST ω),IDENT ω
      MA←'+0×ω'
      MD:ω,(CONST ω)/MA
      DE←'01'
      D:MD ⍎RULE STRIP ω : ∨/END STRIP ω : (END STRIP ω)/DE
      ENP:LP,ω,RP
      D1:(ENP D α),PFS[0],D ω
      D2:PFS[1],D ω
      D3:(ENP D α),PFS[1],D ω
      CL←':'
      DERA:DS,((1+ω⍳CL)↑ω),D (1+ω⍳CL)↓ω
      DERB:BDEF9 DERA ω
      DER:DERB CDEF9 ω
      D14:(ENP (DER α),ω),PFS[2],D ω
      CDEF9:⍎R9,STL9 ω : 3≠⎕NC ω : ODEF9 STL9 ω
      ODEF9:CDEF9 DER 1↓ω : (DS≠1↑ω)∨3≠⎕NC 1↓ω : 0ρω
```

8.3 SOME INTEGRALS

We begin our study of the techniques of integration by collecting the information concerning integrals developed in the previous chapters.

<u>The Plus Rule for Integrals.</u> If *P* ω is equivalent to (*F* ω)+*G* ω and if *IF* and *IG* are integrals of *F* and *G* respectively, then (*IF* ω)+*IG* ω is an integral of *P*.

For example, if *F*:ω and *G*:2○ω then integrals of *F* and *G* are *IF*:0.5×ω*2 and *IG*:1○ω respectively. According to the plus rule for integrals, an integral of ω+2○ω is (0.5×ω*2)+1○ω.

<u>A Scale Rule for Integrals.</u> If *G* ω is equivalent to *F* *A*+*B*×ω for scalars *A* and *B* such that 0≠*B*, and if *IF* is an integral of *F*, then (÷*B*)×*IF* *A*+*B*×ω is an integral of *G*.

For example, if *G*:*1+3×ω and *F*:*ω, then *G* ω is equivalent to *F* 1+3×ω and *IF*:*ω is an integral of *F*. Consequently, according to the scale rule, (÷3)×*1+3×ω is an integral of *G*.

The above rules follows directly from the appropriate rule for derivatives.

Since we already know expressions that define derivatives of many functions we also know expressions which define integrals of many functions. For example, since the ÷1+ω*2 is the derivative of ¯3○ω, it follows that ¯3○ω is an integral of ÷1+ω*2. Integrals of several functions have been collected in the Table 8.3.1 for reference.

	Function	Integral	Restrictions
1.	α**A*	(α**A*+1)÷*A*+1	¯1≠*A*
2.	÷α or α*¯1	⍟|α	
3.	*α	*α	
4.	*B**α	(*B**α)÷⍟*B*	1≠*B*
5.	÷0○α	¯1○α or -¯2○α	
6.	1○α	-2○α	
7.	2○α	1○α	
8.	3○α	-⍟|2○α	
9.	÷4○α	¯5○α	
10.	÷¯4○α	¯6○α	
11.	5○α	6○α	
12.	6○α	5○α	
13.	7○α	⍟6○α	
14.	÷1+α*2	¯3○α	
15.	÷1-α*2	¯7○α	1>|α
	or	0.5×⍟(1+α)÷1-α	
16.	÷1-α*2	0.5×⍟(1+α)÷α-1	1<|α

A TABLE OF INTEGRALS
TABLE 8.3.1

Examples.

	Function	Integral
a.	6×α*2	2×α*3
b.	2○4+α	1○4+α
c.	*5×α	(÷5)×*5×α
d.	÷3+¯4×α	(÷¯4)×⍟|3+¯4×α
e.	2*1-α	-(2*1-α)÷⍟2
f.	*7+α	14+*7+α
g.	(*α)-2×1○α	(*α)+2×2○α

⌹9

The Composition Rule for Derivatives yields a further rule for integrals: suppose that *H ω* is equivalent to (*F G ω*)×*DG ω* and that *IF* is an integral of *F*. Then *IF G ω* is an integral of *H ω*.

Example 1. Suppose that *H*:(*ω)×3+*ω. Then for the functions *F*:3+ω and *G*:*ω, *H ω* is equivalent to (*F G ω*)×*DG ω*. Also, *IF*:(3+ω)+0.5×ω*2 is an integral of *F*. Consequently the function *IF G ω*, or equivalently

(3×*ω)+0.5×(*ω)*2

is an integral of *H*. This can be verified by showing that the derivative of (3×*ω)+0.5×(*ω)*2 is (3×*ω)+(2×0.5×(*ω)*1)×*ω, or equivalently (*ω)×3+*ω.

Example 2. In example 1 we could have defined *F* and *G* as *F*:ω and *G*:3+*ω. Then *H ω* is to (*F G ω*)×*DG ω*, and equivalently *IF*:0.5×ω*2 is an integral of *F*. Consequently *IF G ω*, or 0.5×(3+*ω)*2 is also an integral of *H ω*.

Example 3. If *H*:(*SIN ω*)×*COS ω*, define *F*:ω and *G*:*SIN ω*. Then *H ω* is equivalent to (*F G ω*)×*DG ω* and *IF*:0.5×ω*2 is an integral of *F*. Consequently 0.5×(*SIN ω*)*2 is an integral of *H ω*.

Example 4. In example 3 we could have defined *F* and *G* as *F*:-ω and *G*:*COS ω*. Then *H ω* is equivalent to (*F G ω*)×*DG ω*, and *IF*:-0.5×ω*2 is an integral of *F*. Consequently *IF G ω*, or -0.5×(*COS ω*)*2 is also an integral of *H ω*. ⌹12

8.4 INTEGRATION BY PARTS

Integration by parts is a procedure for producing an integral of the product of two functions *F* and *G*. Evidently we cannot simply form the product of integrals *IF* and *IG*, for according to the Times Rule for Derivatives the derivative of the function of (*IF ω*)×*IG ω* is

((*F ω*)×*IG ω*)+(*IF ω*)×*G ω*

However, consider the function (*IF ω*)×*G ω*, whose derivative is

```
((F ω)×G ω)+(IF ω)×D̲G ω
```

One of the terms (i.e., one of the "parts") of this expression is the product of the functions *F* and *G*; if *H* is the other part, that is if `H:(IF ω)×D̲G ω`, and if *IH* is an integral of *H* then the function

```
((IF ω)×G ω)-IH ω
```

is an integral of `(F ω)×G ω`. To see that this is true, consider the following sequence of identities for the derivative of the above expression:

```
((F ω)×G ω)+(IF ω)×D̲G ω)-H ω
((F ω)×G ω)+((IF ω)×D̲G ω)-(IF ω)×D̲G ω
(F ω)×G ω
```

Thus the following procedure, called <u>integration by parts</u>, can be used to produce integrals of the product of the functions *F* and *G*:

a. choose one of the functions *F* and *G*, say *F*, and produce an integral *IF* of *F*;

b. determine the derivative of the other function, in this case *D̲G*;

c. determine an integral function *IH* of the function `H:(IF ω)×D̲G ω`;

d. the function `((IF ω)×G ω)-IH ω` is an integral of the function `(F ω)×G ω`.

Example 1. To produce an integral of the function `ω×*ω`, define `F:*ω` and `G:ω`. Then `IF:*ω` is an integral of *F* and `D̲G:1+0×ω`. Also,

```
(IF ω)×D̲G ω
*ω
```

Consequently `H ω` is equivalent to `*ω`, and evidently `IH:*ω` is an integral of *H*. Therefore `((IF ω)×G ω)-IH ω`, or equivalently `(ω×*ω)-*ω` is an integral of `ω×*ω`. Clearly the derivative of `(ω×*ω)-*ω` is `(1×*ω)+(ω×*ω)-*ω`, or equivalently `ω×*ω`, as required.

Example 2. To produce an integral of the function `¯1○ω`, define `F:1+0×ω` and `G:¯1○ω`. Then `IF:ω` is an integral of *F* and `D̲G:÷0○ω`. Also,

```
(IF ω)×D̲G ω
ω÷0○ω
```

Consequently `H ω` is equivalent to `ω÷0○ω`, and `IH:-0○ω` is an integral of `H`. Therefore `((IF ω)×G ω)-IH ω`, or equivalently `(ω×¯1○ω)+0○ω` is an integral of `¯1○ω`. Again, the derivative of `(ω×¯1○ω)+0○ω` is `(1×¯1○ω)+(ω÷0○ω)-ω÷0○ω`, or equivalently `¯1○ω` as required.

Example 3. In example 1 we could have reversed the roles of the functions `ω` and `*ω` by defining `F:ω` and `G:*ω`. Then:

```
(IF ω)×D̲G ω
0.5×(ω*2)×*ω
```

and consequently `H ω` is equivalent to `0.5×(ω*2)×*ω`. The problem with this choice of `F` and `G` is that it is just as difficult to produce an integral `IH` of `H` as it is to produce an integral of the original function `ω×*ω`. This choice of `F` and `G` will not lead directly to an integral of `ω×*ω`. The strategy of integration by parts is to choose `F` and `G` in such a way that it is possible to produce an integral `IF` of `F` and an integral of `H:(IF ω)×D̲G ω`, or if that is not possible, to obtain a recursive definition as illustrated in the next example.

Example 4. Integration by parts is sometimes used to produce recursive definitions of integrals of certain functions. For example, consider the function

```
B:(ω*α)×*ω
```

and the problem of producing integrals of the functions `N B ω` for non-negative integers `N`. Define `F:*ω` and `G:ω*α`; then `IF:*ω` is an integral of `F` and `D̲G:α×ω*α-0≠α` is the derivative of `G`, as a function of its right argument. That is, `N D̲G ω` is the derivative of `N G ω` for each constant `N`. Thus, using integration by parts, we see that an integral of `N B ω` is `(IF ω)×N G ω` minus `N` times an integral of `(N-1)B ω`. Also, `0 B ω` is equivalent to `F ω` and so `IF ω` is an integral of `0 B ω`. Consequently, if

```
IB:((IF ω)×α G ω)-α×(α-1)IB ω : 0=α : IF ω
```

then `N IB ω` is an integral of `N B ω` for each non-negative integer `N`. For example, `5 IB ω` is an integral of `(ω*5)×*ω`. ⌹25

8.5 INTEGRATION BY SUBSTITUTION

The <u>substitution method</u> is so called because it produces an integral of `F` by substituting for the arguments of `F` the results of some chosen function `G`. The basic idea is as follows. First of all, we attempt to choose a function `G` for which we can produce an integral of the function `H:(F G ω)×D̲G ω`. Suppose that `IH` is that integral, and that `G` has an inverse function `I̲G`. Then `IH I̲G ω` is an integral of `F ω`. To see that this is true we must show that the derivative of *IH I̲G ω* is identical to *F ω*. According to the definition of *IH*:

```
D̲IH ω                                              [8.5.1]
H ω
(F G ω)×D̲G ω
```

Also, since *G* and *I̲G* are inverse functions, we have from Section 7.1:

```
1                                                  [8.5.2]
(D̲G I̲G ω)×D̲I̲G ω
```

Finally, we have the following sequence of identities for the derivative of *Q:IH I̲G ω*:

```
D̲Q ω
(D̲IH I̲G ω)×D̲I̲G ω                     Composition Rule
(F G I̲G ω)×(D̲G I̲G ω)×D̲I̲G ω          Identity 8.5.1
(F G I̲G ω)×1                          Identity 8.5.2
F ω                                    Definition of I̲G
```

Example 1. Suppose that *F:ω÷(1+ω)*0.5*. Choose *G* to be *G:¯1+ω*2*. Then *D̲G:2×ω* and

```
(F G ω)×D̲G ω
((¯1+ω*2)÷ω)×2×ω
2×¯1+ω*2
```

Evidently the function *IH:2×(-ω)+(ω*3)÷3* is an integral of *H:(F G ω)×D̲G ω*. The function *I̲G:(1+ω)*0.5* is the inverse function of *G*. Therefore:

```
IH I̲G ω
2×(-(1+ω)*0.5)+((1+ω)*1.5)÷3
```

and the latter expression is an integral of *F*. To see that this is true, the derivative of *2×(-(1+ω)*0.5)+((1+ω)*1.5)÷3* is *1×(-(1+ω)*¯0.5)+(1+ω)*0.5*, or equivalently *ω÷(1+ω)*0.5*. The latter expression is the definition of *F*.

Note that *G* is chosen to be the inverse function of the denominator *(1+ω)*0.5* of *F*, so that upon substituting *G ω* for *ω* in this denominator, the result is the simple expression *ω*. The general strategy in applying the substitution method is to choose *G* in such a way as to simplify one of the factors of *F*.

Example 2. Suppose that *F:(9-ω*2)*0.5*. Choose *G* to be *G:3×1○ω*, where *(|ω)≤○÷2*. Then

```
(F G ω)×D̲G ω
((9-9×(1○ω)*2)*0.5)×3×2○ω
9×(2○ω)*2
```

Since

```
9×(2○ω)*2
4.5+4.5×2○2×ω
```

Exercise 8.13

the function *IH*:(4.5×ω)+2.25×1○2×ω is an integral of *H*:(*F G* ω)×*D̲G* ω. The function *I̲G*:¯1○ω÷3 is the inverse function of *G*, and *IH I̲G* ω is an integral of *F* ω.

Note that *G* is chosen to take advantage of the relation 1=+/(1 2○ω)*2, the result being that *F G* ω is simply 2×2○ω.

Example 3. Suppose that *F*:(ω*5)÷4○ω. Choose *G*:¯4○ω. Then *D̲G*:ω÷¯4○ω and

```
(F G ω)×D̲G ω
(((¯4○ω)*5)÷ω)×ω÷¯4○ω
(¯4○ω)*4
(¯1+ω*2)*2
1+(¯2×ω*2)+ω*4
```

Evidently *IH*:ω+((¯2÷3)×ω*3)+(÷5)×ω*5 is an integral of *H*:(*F G* ω)×*D̲G* ω. Since *I̲G*:4○ω is the inverse function of *G* then *IH* 4○ω is an integral of *F*. ⊞27

8.6 INTEGRATION OF RATIONAL FUNCTIONS

For vectors *B* and *C*, the function defined by (*B POLY* ω)÷*C POLY* ω is called a rational function. Integrals of rational functions can be produced by way of the partial fraction decomposition of rational functions.

For example, for the function *F*:2×ω÷(1+ω)×1-ω we have the identity

```
F ω
(÷1-ω)-÷1+ω
```

The latter expression is the partial fraction decomposition of *F* ω. Note that it is easy to produce an integral of *F* from the partial fraction decomposition. Namely, (⍟|1-ω)-⍟|1+ω is an integral of *F* ω.

In order to produce the partial fraction decomposition of *F* ω, we must determine the scalars *A* and *B* for which

```
F ω
(A÷ω-1)+B÷ω+1
```

or equivalently

```
F ω
(÷ω∘.-1 ¯1)+.×A,B
```

Thus if `X` is a vector of length 2 for which `0≠DET ÷X∘.-1 ¯1` then the preceding identity yields

```
      A,B
      (F X)⌹÷X∘.-1 ¯1
```

and the latter expression will produce the vector `A,B`:

```
      X←4 5
      0≠DET ÷X∘.-1 ¯1
1

      (F X)⌹÷X∘.-1 ¯1
¯1 ¯1
```

As a second example, for the function `G:(ω+1)÷(ω-2)*2` we have the identity

```
      G ω
      (÷ω-2)+3÷(ω-2)*2
```

The latter expression is the partial fraction decomposition of `G ω`, and an integral of `G ω` is, evidently, `(⍟|ω-2)+¯3÷ω-2`. In order to produce the partial fraction decomposition of `G ω`, we must determine scalars `A` and `B` for which

```
      G ω
      (A÷ω-2)+B÷(ω-2)*2
```

or equivalently

```
      G ω
      (÷(ω∘.-2 2)*((ρω),2)ρ1 2)+.×A,B
```

Thus if `X` is a vector of length 2 for which

```
      0≠DET ÷(X∘.-2 2)*2 2ρ1 2
```

then the preceding identity yields

```
      A,B
      (G X)⌹÷(X∘.-2 2)*2 2ρ1 2
```

and the latter expression will produce the vector `A,B`:

```
      X←0 1
      0≠DET ÷(X∘.-2 2)*2 2ρ1 2
1
      (G X)⌹÷(X∘.-2 2)*2 2ρ1 2
1 3
```

In general, suppose that `C POLY ω` has no complex roots. Then there is a lean vector `R0` and a vector of positive integers *E*0 for which

```
      C POLY ω                                    [8.6.1]
      (ω∘.-R0)×.*E0
```

The integer *E*0[*I*] is called the multiplicity of the root *R*0[*I*]; *R*0[*I*] is said to be a simple root if *E*0[*I*] is 1, and a multiple root if *E*0[*I*] is greater than 1. If

```
      MR:(α[0]ρω[0]),(1↓α) MR ¯1↓ω : 0=ρα : ι0
```

(for multiple roots) and *R*←*E*0 *MR* *R*0, then each element *R*0[*I*] appears *E*0[*I*] times in *R*. The length of *R* is *DEGREE C* and 8.6.1 is equivalent to

```
      C POLY ω                                    [8.6.2]
      ×/ω∘.-R
```

For example:

```
      R0←1 ¯4 2 5
      E0←2 3 1 1
      R←E0 MR R0
      R
1 1 ¯3 ¯3 ¯3 2 5
```

We now define

```
      CT:(1+ια[0]),CT 1↓α : 0=ρα : ι0
```

(for count) and *E*←*CT* *E*0. For example:

```
      E0
2 3 1 1
      E←CT E0
      E
1 2 1 2 3 1 1
```

Finally, we define

```
      RC:((ρα),ρω)ρω
```

(for row copies). For example:

```
      0 ¯1 2 RC 3 4 ¯6 0 1
3  4 ¯6  0  1
3  4 ¯5  0  1
3  4 ¯6  0  1
```

Then if *DEGREE B* is less than *DEGREE C*, there is a vector *A* of length *DEGREE C* for which

```
      (B POLY ω)÷C POLY ω                         [8.6.3]
      (÷(ω∘.-R)*ω RC E)+.×A
```

The latter expression is the partial fraction decomposition of the rational function

```
(B POLY ω)÷C POLY ω
```

If *X* is a vector of length *DEGREE C* for which

```
0≠DET ÷(X∘.-R)*X RC E
```

then the vector *A* in Identity 8.6.3 is produced by the following expression, where *F*:(*B POLY* ω)÷*C POLY* ω:

```
(F X)⌹÷(X∘.-R)*X RC E                          [8.6.4]
```

For example, the elements of

```
R0←3 5 ¯1
```

are the roots of ¯45 ¯6 28 ¯10 1 *POLY* ω with the corresponding multiplicity vector *E*0←2 1 1. Suppose that *B*←¯41 49 ¯7 1 and define *R*←*E*0 *MR R*0 and *E*←*CT E*0. If *X*←0 1 ¯2 2 and *M*←÷(*X*∘.-*R*)*(⍴*X*) *RC E* then 0≠*DET M* Thus we define *C*←¯45 ¯6 28 ¯10 1, and

```
      A←((B POLY X)÷C POLY X)⌹M
      A
2 ¯2 ¯4 1
```

Therefore the rational function

```
(B POLY ω)÷C POLY ω
```

is equivalent to 2÷ω-3 plus ¯2÷(ω-3)*2 plus ¯4÷ω-5 plus 1÷ω-¯1. It follows that an integral of this rational function is 2×⍟|ω-3 plus 2÷ω-3 plus ¯4×⍟|ω-5 plus ⍟|ω+1.

Note that if (*DEGREE B*)≥*DEGREE C* then according to the division algorithm for polynomials (Ex 3.36) there exist coefficient vectors *Q* and *RM* such that (*DEGREE RM*)<*DEGREE C* and

```
(B POLY ω)÷C POLY ω
(Q POLY ω)+(RM POLY ω)÷C POLY ω
```

Thus to produce an integral of the rational function defined by the first expression in this identity, we apply the above procedure to the rational function (*RM POLY* ω)÷*C POLY* ω and add to the result an integral of *Q POLY* ω. ⌹31

Exercises

0.1 Evaluate each of the following:

```
3*3                 1E2=10E1
4⌈5                 |¯15E5
3⌊3                 ÷4
1.2≥1               ÷0.5
×¯4                 1>×1
```

0.2 Evaluate each of the following:

```
3+4*2               (1∧1)∨0
(3+4)*2             (4-7)×5+2
                         ¯2×1≠4            (4-7)×(5+2)
1∧1∨0               1=4≠1E8
```

0.3 Let *A*←0 4 5 and *B*←2 1 2. Evaluate

```
A+B                 A*B
B×A                 B-A
A⌈B
```

There is a convenient scheme for producing tables such as that defined by *A*∘.+*B*. Define *A*←1 2 and *B*← ¯3 0 4. Write down *A* and *B* and fill in the table as indicated below.

```
      |¯3 0 4 B
A    1|¯2 1 5
     2|¯1 2 6
```

0.4 Use the scheme illustrated above to produce the table *A*∘.+*B* for *A*←1 ¯3 1.5 and *B*←¯4 1.

0.5 Use the scheme illustrated above to produce the tables *A*∘.×*B* and *A*∘.⌈*B* for *A*←1 4 7 and *B*←3 ¯3 5 1 4.

0.6 Repeat exercise 3 for *A*←3 ¯2 1 and *B*←3.

0.7 Evaluate

```
4↑7 8 9 3 ¯1        8↑7 8 9 3 ¯1
¯1↓7 8 9 3 ¯1       ¯8↑7 8 9 3 ¯1
```

```
1 0 1 1 0/4 7 8 2 9
1 0 0 0 1/4 7 8 2 9
1 0 1 1 0\1 0 1 1 0/4 7 8 2 9
1 1 0 0 1/1 1 0 0 1\6 5 3

A←1 0 3 1∘.×¯4 1 2 3

3 2↑A                   ¯2 ¯2↑A
3 2↓A                   ¯2 ¯2↓A
1 0↓A                   ¯2 0↓A
0 1↓A                   0 ¯2↓A
```

0.8 a. Let *X*←7 3 ¯2 4 6 5 and evaluate

```
¯2↑X                    (¯4↑X)[0 2]
X[0 3 5]                ¯6↑¯4↓X
4↓X                     6↑¯4↓X
(4↑X)[0 2]              X[5 4 3 2 1 0]
```

b. Let *M*←1 0 ¯1∘.+5 3 1 4 and evaluate

```
M[1;1]        M[2;]        M[;2]
M[0;3]        M[0;0]       M[2;3]
```

0.9 Let *A*←4 7 ¯2 and *B*←2 0 7 ¯3 and evaluate

```
A,B                     (B,A)[0 4 1]
(7↑A)+¯7↑B              (A,B)[0 4 1]
-B,-A                   3↓A,B
(-B),A                  A,A,B,A
```

0.10 Evaluate each of the following:

```
+\5 ¯8 1 0 0            0=-/2 4 1 0 1
×\5 ¯8 1 0 0            -\2 4 1 0 1
⌈\5 ¯8 1 0 0            ⌊\5 4 3 2 7
```

0.11 Let *M*←1 ¯2∘.×4 7 3 and evaluate +/*M* and +⌿*M*.

0.12 Let *N*←2 0 ¯3∘.+1 5 7 and evaluate ×/*N* and ×⌿*N*.

0.13 Evaluate each of the following:

```
⍴1 2 ¯5 7                        ⍴⍳9
(⍴1 0 1)=⍴2 5 10                 ⍴⍳100
⍴2 5 10∘.÷1 3 7 9                (10⍴1 0)/⍳10
(⍴1 0∘.=0 1)=⍴4 6∘.+5 9 ¯7       (10⍴1 1 0)/⍳10
```

0.14 Show that the value of each of the following expressions is 1.

```
14=+/7ρ2                256=×/9ρ1 4
15=+/5ρ3                256≠×/10ρ1 4
256=×/8ρ1 4
```

0.15 Evaluate `×/4 4ρι16` and `×⌿4 4ρι16`.

0.16 Evaluate each of the following:

```
⌽7 9 4 2 8              +/⌽5 ¯3 4
2⌽7 9 4 2 8             +/1⌽5 ¯3 4
¯3⌽7 9 4 2 8            +/¯2⌽5 ¯3 4
```

0.17 Test each of the following identities for various choices of the vectors *F* and *G*:

```
+/F,G                   F[0]××/1↓F
(+/F)++/G               ×/F
```

0.18 Evaluate `ι3` and `ι2` and `ι1`. `ι0` is also a vector, and by analogy with the values of `ι3` and `ι2` and `ι1`, `ι0` has no elements. `ι0` is called an <u>empty</u> vector. Even though `ι0` has no elements it is possible to use this vector as a function argument. For example, `ι0` is a valid argument of the function called catenate, and evidently

```
V,ι0
V
```

for vectors *V*. `ι0` is also a valid argument of the function `+/ω`. What value for `+/ι0` is suggested by the following sequence of identities?

```
+/V
+/V,ι0
(+/V)++/ι0
```

0.19 Selection functions <u>distribute</u> <u>over</u> scalar functions. For example, we have the identity

```
V[0 2 3]+W[0 2 3]
(V+W)[0 2 3]
```

for vectors *V* and *W* whose sizes are at least `4`, that is, `(4≤ρV)∧4≤ρW`. Test this identity.

0.20 Test the identities

```
(1↓X)×1↓Y               (⌽A)⌈⌽B
1↓X×Y                   ⌽A⌈B
```

0.21 a. Without using recursion define the function *F* whose value *F* 0 is 1 and *F N* is *N*×*F N*-1 for a positive integer *N*.

b. Define the function *G* for which *G X* is ¯2 if *X* is negative, 1 if *X* is 0, and 3 if *X* is positive.

c. Define the function *H* such that *H Y* is the vector of the positive elements of the vector *Y*.

1

1.1 There is a convenient scheme for producing values of the function *DIFF*. For example, the vector *DIFF* 2 5 10 14 can be produced as follows:

```
   2 5 10 14
 -   2  5 10 14
   ------------
     3  5  4
```

Use this scheme to produce the vector *DIFF Y* for each vector *Y* below.

```
a. Y←10 4 6 8 3
b. Y←0 ¯5 8 ¯7
```

1.2 For the vector *F*←10 8 6 5 3 7 11 we have

```
      DIFF F
¯2 ¯2 ¯1 ¯2 4 4
```

The first 4 elements of *DIFF F* are negative, indicating that the first 5 elements of *F* are decreasing in value. Also, the last 2 elements of *DIFF F* are positive, indicating that the last 3 elements of *F* are increasing in value. Since the values of the elements of *F* decrease until *F*[4] and then increase, we must have *F*[4]=⌊/*F*.

Now suppose that *F* is a vector for which

```
      DIFF F
¯3 ¯7 ¯2 1 4 1 2
```

For which element *F*[*I*] is *F*[*I*]=⌊/*F*?

1.3 Suppose that *G* is a vector for which

```
      DIFF G
7 2 1 ¯5 ¯1 ¯1 ¯2
```

For which element *G*[*J*] is *G*[*J*]=⌈/*G*?

1.4   For the vector F in Ex 1.2, can the element $F[K]$ for which $F[K]=\lceil/F$ be determined? If so, what is K?

1.5   For the vector G in Ex 1.3, can the element $G[K]$ for which $G[K]=\lfloor/G$ be determined? If so, what is K?

1.6   Show that:

> *DIFF Y*   *DIFF Y*
> $1\downarrow Y-\mbox{}^{-}1\phi Y$ $\mbox{}^{-}1\downarrow(1\phi Y)-Y$

Hint.   Test these identities with schemes similar to the one in Ex 1.1.   These tests may suggest proofs.

———————

1.7   Show that for vectors V

> $(+\backslash V)[I+1]=(+\backslash V)[I]+V[I+1]$

by verifying the sequence of identities:

> $(+\backslash V)[I+1]$
> $+/V[\iota I+2]$
> $(+/V[\iota I+1])+V[I+1]$
> $(+\backslash V)[I]+V[I+1]$

1.8   Ex 1.7 suggests a simple scheme for producing the vector $+\backslash V$ for a vector V. Namely, once the element $(+\backslash V)[I]$ has been determined then the element $(+\backslash V)[I+1]$ is the sum of $(+\backslash V)[I]$ and $V[I+1]$. For example, the vector $+\backslash 2\ 5\ 9\ \mbox{}^{-}3\ 4$ can be produced as follows:

```
2 5  9 ⁻3  4
2
2 7
2 7 16
2 7 16 13
2 7 16 13 17
```

Determine the vector $+\backslash V$ for each vector V below.

> a. $V\leftarrow0\ 3\ 0\ \mbox{}^{-}5\ 16$ b. $V\leftarrow1\ 2\ \mbox{}^{-}1\ 4\ 8$

1.9   Use the definition of the function *SUM* and the computational scheme in Ex 1.8 to produce each of the following vectors.

> a. $3+SUM\ \mbox{}^{-}2\ 1\ 7\ \mbox{}^{-}9\ 4$ c. $\mbox{}^{-}1+SUM\ 0\ 2\ 5\ \mbox{}^{-}8\ 3$
> b. $0+SUM\ 4\ 11\ 17\ 3$

———————

1.10   Test the identities 1.1.1 for each pair of a scalar C and a vector F below.

> a. $C\leftarrow5\ ;\ F\leftarrow4\ \mbox{}^{-}7\ 9\ 0$ b. $C\leftarrow\mbox{}^{-}3\ ;\ F\leftarrow3\ 0\ 1\ 2\ 0\ \mbox{}^{-}5$

1.11 Use the function *SUM* to produce the vector *F* described in Ex 1.2 for which `F[0]=3`. Check your solution to Ex 1.2.

1.12 Repeat Ex 1.11 for the vector *F* for which `F[0]=¯5`.

1.13 Repeat Ex 1.4 for the vector *F* determined in Ex 1.11.

1.14 Repeat Ex 1.11 for the vector *G* described in Ex 1.3 for which `G[0]=0`.

1.15 Repeat Ex 1.5 for the vector *G* determined in Ex 1.14.

Exs 1.16-19 are concerned with the proofs of the Identities 1.1.1. You may find it helpful to illuminate the sequences of identities with several examples before verifying them. For instance, you might work through the identities in Ex 1.16 with the vector `V←2 5 ¯1 8`.

1.16 Show that for vectors *V*

```
V
V[0]+SUM DIFF V
```

by verifying the sequence of identities:

```
(V[0]+SUM DIFF V)[0]
(V[0]+0,+\DIFF V)[0]
V[0]+0
V[0]
```

and for *I* such that `0<I`:

```
(V[0]+SUM DIFF V)[I]
(V[0]+0,+\DIFF V)[I]
V[0]+(+\DIFF V)[I-1]
V[0]++/(DIFF V)[⍳I]
V[0]++/(1↓V[⍳I+1])-¯1↓V[⍳I+1]
V[0]+V[I]-V[0]
V[I]
```

1.17 Show that for vectors *X* and *Y*

```
DIFF X+Y
(DIFF X)+DIFF Y
```

by verifying the sequence of identities:

```
DIFF X+Y
(1↓X+Y)-¯1↓X+Y
((1↓X)+1↓Y)-(¯1↓X)+¯1↓Y
((1↓X)-¯1↓X)+(1↓Y)-¯1↓Y
(DIFF X)+DIFF Y
```

1.18 Show that for scalars *C*

```
DIFF NρC
(N-1)ρ0
```

1.19 Show that for vectors *F* and scalars *C*

```
F
DIFF C+SUM F
```

by verifying the sequence of identities:

```
(DIFF C+SUM F)[I]
(DIFF C+0,+\F)[I]
(DIFF ((1+ρF)ρC)+0,+\F)[I]
((DIFF (1+ρF)ρC)+DIFF 0,+\F)[I]
(0+DIFF 0,+\F)[I]
(DIFF 0,+\F)[I]
(0,+\F)[I+1]-(0,+\F)[I]
(0,+\F)[I]+F[I]-(0,+\F)[I]
F[I]
```

1.20 Show that `∧/6=DIFF Y` for each of the vectors *Y* below:

```
Y←6×ι5                    Y←6×7+ι8
Y←3+6×ι5                  Y←¯8.3+6×3+ι4
```

1.21 Show that `∧/4=DIFF DIFF Z` for each of the vectors *Z* below:

```
Z←2×(ι6)*2
Z←7+2×(ι7)*2
Z←¯2.5+2×1+(ι5)*2
```

1.22 Use the function *SUM* to determine *Y* such that

```
(3=Y[0])∧∧/2 2 2 2=DIFF Y
```

1.23 Determine the vector *Z* for which

```
(¯5=Z[0])∧∧/10 10 10 10 10=DIFF Z
```

1.24 Determine the vector *W* for which

```
(0=W[0])∧(4=(DIFF W)[0])∧∧/(4ρ15)= DIFF DIFF W
```

1.25 Determine the vector *W* for which

```
(5=W[0])∧(¯5=(DIFF W)[0])∧∧/(6ρ7)=DIFF DIFF W
```

1.26 For the vector `X←2 5 ¯3 7` determine *Y* such that

```
(5=Y[0])∧∧/(3ρ2)=(DIFF Y)÷DIFF X
```

Hint. The equations `∧/(3ρ2)=(DIFF Y)÷DIFF X` and `∧/(DIFF Y)=(3ρ2)×DIFF X` are equivalent. Use the procedure in Exs 1.22 and 1.23 to determine *Y* such that `(5=Y[0])∧∧/(DIFF Y)=(3ρ2)×DIFF X`.

1.27 For the vector `U←1 3 6 10 15` determine *V* such that

```
(0=V[0])∧∧/(4ρ¯3)=(DIFF V)÷DIFF U
```

1.28 For the vector `A←1 9 13 21` determine *T* such that

```
(2=T[0])∧∧/(3ρ4)=(DIFF A)÷DIFF T
```

1.29 The following lunar landing problem provides an interesting application of *SUM* and *DIFF*. Suppose that we begin to track a lunar landing vehicle when it is at some initial altitude `A0` above the lunar surface and continue to track the vehicle at one second intervals during its descent. The altitudes of the vehicle can be represented by a vector *A* for which `A0=A[0]` and `A[I]` represents the altitude in feet of the vehicle *I* seconds after the initial tracking.

For each index *I* the scalar defined by the expression `(A[I+1]-A[I]÷(I+1)-I`, or equivalently `A[I+1]-A[I]`, represents an average velocity of the landing vehicle as it moves from the altitude `A[I]` to `A[I+1]`. The vector `V←DIFF A` is the vector of velocities. The scalar defined by the expression `V[I+1]-V[I]` is an average acceleration of the vehicle as it moves from the altitude `A[I]` to `A[I+2]`. The vector `DIFF V`, or equivalently `DIFF DIFF A`, is called the vector of accelerations.

If the landing vehicle is close to the lunar surface then the acceleration imparted to the vehicle by lunar gravity does not depend on the altitude of the vehicle and is approximately `¯6` feet per second per second. (The sign of the acceleration is negative to indicate that lunar gravity pulls the vehicle towards the lunar surface. Similarly, as the vehicle descends the sign of its velocity is negative; if the vehicle ascends the sign of its velocity is positive). If the vehicle falls freely then:

```
DIFF DIFF A
(¯2+ρA)ρ¯6
```

A vector of altitudes *A* can be produced from the preceding identity using the function *SUM*. If the first two elements of *A* are known then the first element of the velocity vector *V* is also known (namely, `V[0]=A[1]-A[0]`), and this is enough information to determine the vector *A*. For example, if the initial altitude of the vehicle is 10000 feet and the altitude 1 second later is 9250 feet, then the first element of the velocity vector has the value `¯750`. The subsequent altitudes can be determined from the functions *VEL*:α+*SUM* ω and *ALT*:α+*SUM* ω as follows:

```
V←¯750 VEL 13⍴¯6

A←10000 ALT V
```

A	V
10000	¯750
9250	¯756
8494	¯762
7732	¯768
6964	¯774
6190	¯780
5410	¯786
4624	¯792
3832	¯798
3034	¯804
2230	¯810
1420	¯816
604	¯822
¯218	¯828
¯1046	

Since *A*[12]=604 and *A*[13]=¯218, the vehicle landed between 12 and 13 seconds after the initial tracking. Its velocity at impact was between ¯822 and ¯828 feet per second.

In the preceding example the landing velocity was approximately ¯560 miles per hour. Obviously the vehicle would not be allowed to fall freely; it would be equipped with a rocket to slow its descent. Suppose that this rocket can be fired in bursts of 1 second duration and upon firing it imparts an acceleration of 50 feet per second per second to the vehicle. Thus whenever the rocket is firing the total acceleration of the vehicle is 50-6 feet per second per second. For example, for *ACC*:¯6+50×ω, if the rocket is fired every other second starting with the first of a total of 10 seconds, then the accelerations during this 10 second interval are *ACC* 10⍴1 0:

```
      ACC 10⍴1 0
44 ¯6 44 ¯6 44 ¯6 44 ¯6 44 ¯6
```

The pilot's strategy for firing the rocket determines a vector of accelerations from which the resulting altitudes can be determined as in the first example. For example, suppose that, as in the first example, the initial altitude of the vehicle is 10000 feet and the altitude 1 second later is 9250 feet. If the pilot fires the rocket for 10 seconds and then turns it off for 5 seconds:

```
V←¯750 VEL ACC 15↑10⍴1

A←10000 ALT V
```

```
      A              V         ACC 15↑10ρ1
  10000          ¯750             44
   9250          ¯706             44
   8544          ¯662             44
   7264          ¯574             44
   6690          ¯530             44
   6160          ¯486             44
   5674          ¯442             44
   5232          ¯398             44
   4834          ¯354             44
   4480          ¯310             ¯6
   4170          ¯316             ¯6
   3854          ¯322             ¯6
   3532          ¯328             ¯6
   3204          ¯334             ¯6
   2870          ¯340
   2530
```

At the end of this firing interval the pilot must determine a new firing strategy; Suppose that the rocket is to be fired every other second starting with the first of a total of 15 seconds. In order to construct the next altitude vector two initial altitudes are needed. The obvious choice is the last two elements of the previous altitude vector.

```
      V←(-/A[16 15]) VEL ACC 15ρ1 0

      A←A[15] ALT 15ρ1 0

      A              V         ACC 15ρ1 0

   2870          ¯340             44
   2530          ¯296             ¯6
   2234          ¯302             44
   1932          ¯258             ¯6
   1674          ¯264             44
   1410          ¯220             ¯6
   1190          ¯226             44
    964          ¯182             ¯6
    782          ¯188             44
    594          ¯144             ¯6
    450          ¯150             44
    300          ¯106             ¯6
    194          ¯112             44
     82           ¯68             ¯6
     14           ¯74             44
    ¯60           ¯30
    ¯90
```

The vehicle has landed at approximately ¯74 feet per second.

In the previous example the landing velocity is approximately ¯50 miles per hour, a definite improvement over falling freely but still a bone cruncher. The lunar landing problem is to determine a strategy for firing the retro rocket so that the vehicle lands at a velocity of less than, say, ¯25 feet per second (starting at the initial altitudes of 10000 feet and 9250 feet). Also, you have only a limited amount of fuel with which to work, say enough for 25 seconds of firing time. Try it.

1.30 There is a convenient computational scheme for experimenting with the lunar landing problem. For example, suppose that the vector of accelerations is *AC*←¯6 44 ¯6 44 ¯6 and that the first two altitudes of the landing vehicle are 1500 feet and 1250 feet. Consequently the first element of the vector of velocities is ¯250. This information can be organized in a table as follows:

AC			¯6	44	¯6	44	¯6
V		¯250					
A	1500	1250					

Since *AC*=*DIFF V* it follows that *V*[*I*+1]=*V*[*I*]+*AC*[*I*]. Similarly, since *V*=*DIFF A* it follows that *A*[*I*+1]=*A*[*I*]+*V*[*I*]. Thus the missing elements of the above table can be filled in by first producing the element *V*[1], then the element *A*[2], then *V*[2], followed by *A*[3], and so on. As you proceed to fill in the table in this order you can estimate whether the vehicle is falling too fast or too slow and adjust the remaining elements of the vector *AC* accordingly. The following illustration shows the completed table.

AC			¯6	44	¯6	44	¯6
V		¯250	¯256	¯212	¯218	¯174	¯130
A	1500	1250	994	782	564	390	260

Use this computational scheme to experiment with the lunar landing problem.

1.31 To determine a vector *Y* for which ∧/(*DIFF Y*)=¯1↓*R*×*Y* for a scalar *R*, consider the following sequence of identities:

```
∧/(DIFF Y)=¯1↓R×Y
∧/((1↓Y)-¯1↓Y)=¯1↓R×Y
∧/(1↓Y)=(¯1↓Y)×1+R
```

The last equation in this sequence can be written element-by-element as follows. The *I*th element of the vector 1↓*Y* is identical to *Y*[*I*+1] and the *I*th element of the vector (¯1↓*Y*)×1+*R* is identical to *Y*[*I*]×1+*R*. Consequently:

```
Y[I+1]=Y[I]×1+R
```

Thus if the element *Y*[0] is known the vector *Y* can be produced element-by-element, starting with *Y*[1]. For example, to determine the vector *Y* for which

```
(5=ρY)∧(6=Y[0])∧(DIFF Y)=¯1↓2×Y
```

we proceed as follows:

```
      Y←1ρ6
      Y←Y,[0]×1+2
      Y←Y,[1]×1+2
      Y←Y,Y[2]×1+2
      Y←Y,Y[3]×1+2
      Y←Y,Y[1]×1+2

      Y                        DIFF Y                    ¯1↓2×Y
6 18 54 162 486          12 36 108 324           12 36 108 324
```

a. Determine the vector *Y* for which

```
(7=ρY)∧(1=Y[0])∧∧/(DIFF Y)=¯1↓3×Y
```

b. Plot the values of *Y* against the arguments ιρ*Y*.

1.32 Repeat Ex 1.31 for the vector *V* for which

```
(9=ρV)∧(2=V[0])∧∧/(DIFF V)=¯1↓V
```

1.33 It was projected in 1970 that the population of the United States would increase by 2 per cent each year until 1980. If the 1970 population is 2*E*8, what is the projected population for the year 1980?
Hint. Let *P* denote the vector whose *I*th element is identical to the population in the year 1970+*I*. To say that the population will increase by two per cent each year is providing you with information about the rate at which the population is changing. This rate of change is expressed by (*DIFF P*)÷*DIFF* ιρ*P*, or equivalently, *DIFF P*. Two per cent of the population in the year 1970+*I* is defined by the expression 0.02×*P*[*I*]. Thus the statement "the population will increase by 2 per cent each year" is a statement, in words, of the equation

```
∧/(DIFF P)=¯1↓0.02×P
```

Now use the procedure illustrated in Ex 1.31 to produce the vector *P*.

1.34 It is also projected that the population of the United States will decrease by 1 per cent each year from 1980 to 2000. If the population in 1980 is 2.44*E*8, what is the projected population for the year 2000?

1.35 The purpose of this exercise is to illustrate how our knowledge of a function's behavior increases as we compare pairs of function values for pairs of neighboring arguments. The graph of each function defined below is to be produced on the interval (¯5≤ω)∧ω≤5. The strategy to be used is as follows:

Step A. Choose two integer arguments, for example ¯3 and 4, and two other arguments close to these two, for example ¯2.9 and 4.2. Plot the four points on the graph of the functions with those four arguments. Sketch two segments of the graph near the two pairs of neighboring points as in the example in the text. Try to sketch the entire graph using these two arguments.

Step B. Choose a new integer argument, for example 2, and plot the point on the graph of the function with this argument. Try to sketch the entire graph using this new point and the segments already graphed.

Step C. Choose a new argument near the integer argument chosen in step B, for example 2.1. Plot the point on the graph of the function with this argument and sketch a segment of the graph of the function near this new pair of neighboring points. Try to sketch the entire graph.

Step D. Repeat steps B and C until all the integer arguments between ¯5 and 5 have been used.

The functions to be graphed:

a. `¯3+(¯2×ω)+ω*2`
b. `¯4+(5×ω)-ω*2`
c. `1+(¯12×ω)+(3×ω*2)+2×ω*3`
d. `5+(3×ω)+ω*3`
e. `×/ω-¯1 1 2`

1.36 Use the graphing technique of Ex 1.35 to sketch the graph of each of the following functions.

a. `÷1+ω`
b. `÷(1+ω)×1-ω`
c. `3÷(2-ω)*2`
d. `ω÷(1+ω)×(1-ω)*2`

Hint. Each function in this exercise is of the form `(F ω)÷G ω`. It will be helpful to first sketch each graph for arguments near those *X* for which `0=G X`.

1.37 Even though the values `0 QF X` of a difference-quotient function are indeterminate, the values `S QF X` often approach a unique value as *S* is assigned values closer and closer to zero. Test this fact by defining the difference-quotient function *QF* of each of the following functions and evaluating `S QF X` for the indicated value of *X* and `S←(0.1*⍳7)∘.×1 ¯1`.

a. `F:ω*3 ;X←2`
b. `F:ω+ω*2 ;X←1`
c. `F:ω÷ω-1 ;X←¯1`
d. `F:1+ω*4 ;X←0`

1.38 Repeat Ex 1.37 for each of the following functions, and carefully describe your observations.

a. `F:×ω-3 ;X←3`
b. `F:1+|ω-4 ;X←4`

1.39 Binomial coefficients can be produced by the primary function α!ω. The monadic function !ω is called <u>factorial</u>; if *N* is a non-negative integer then !*N* is ×/1+ι*N*. If *M* is also a non-negative integer and is less than or equal to *N*, then *M*!*N* is (!*N*)÷(!*M*)×!*N*-*M*. Test the identity

```
BIN N
(ιN+1)!N
```

1.40 Tables of binomial coefficients can be produced by the expression (ι*N*)∘.!ι*N*. For example:

```
      0⍕T←(ι10)∘.!ι10
 1  1  1  1  1  1  1  1  1   1
 0  1  2  3  4  5  6  7  8   9
 0  0  1  3  6 10 15 21 28  36
 0  0  0  1  4 10 20 35 56  84
 0  0  0  0  1  5 15 35 70 126
 0  0  0  0  0  1  6 21 56 126
 0  0  0  0  0  0  1  7 28  84
 0  0  0  0  0  0  0  1  8  36
 0  0  0  0  0  0  0  0  1   9
 0  0  0  0  0  0  0  0  0   1
```

Vectors of binomial coefficients appear along the columns of this table. The vectors along the rows are called vectors of <u>figurate numbers</u>. Test the following identity for vectors of figurate numbers:

```
SUM ¯1↓T[I;]
T[I+1;]
```

1.41 The Binomial Theorem suggests a procedure for evaluating the functions ω⋆*N*. For example, 4.1⋆3 is equal to (4+0.1)⋆3, and according to the Binomial Theorem

```
(4+0.1)⋆3
+/(4⋆ι4)×1 3 3 1×0.1⋆⌽ι4
```

Write down the vectors 4⋆ι4, 1 3 3 1, and 0.1⋆⌽ι4 in columns as follows:

```
 1   1    .001
 4   3    .01
16   3    .1
64   1   1
```

Form a new column by taking the product over the rows in this table, and then sum the elements of the new column.

```
 1   1    .001       .001
 4   3    .01        .12
16   3    .1        4.8
64   1   1         64
                 ________
                   68.921
```

Use the procedure illustrated above to evaluate each of the following expressions. The appropriate vectors of binomial coefficients can be found in the table *T* of Ex 1.40.

a. 3.1*4
b. 2.02*3
c. 2.5*3
d. 3.9*4

1.42 The Binomial Theorem also suggests a procedure for producing <u>approximate</u> values of the functions *ω*N*. For example, consider again 4.1*3 and the four column table of Ex 1.41 with the sum over the last column deleted.

1	1	.001	.001
4	3	.01	.12
16	3	.1	4.8
64	1	1	64

Summing only the last three elements in the last column produces an approximate value of 4.1*3 which is accurate to two decimal places, and summing the last two elements produces a result which is accurate for the integer part only. Use this procedure to produce approximate values of the expressions below which are accurate to one decimal place.

a. 2.1*5
b. 2.02*4
c. 5.5*3
d. ¯4.99*4

1.43 Repeat Ex 1.42 to produce approximate values which are accurate to two decimal places.

1.44 Show that the cline function of *P*0:*ω**0 is *CP*0:0×*α*+*ω* and the cline of *P*1:*ω**1 is *CP*1:1+0×*α*+*ω*. Note that the expression 0×*α*+*ω* in the definition of *P*0 produces a dyadic function whose value if 0 for all arguments.

1.45 Use the Binomial Theorem to define the cline functions of *P*2:*ω**2, *P*3:*ω**3, *P*4:*ω**4, and *P*5:*ω**5.

1.46 Use the definition of the cline function *CP*0 in Exercise 1.44 to prove 1.3.4 in the case when *N* is 0.

1.47 Use the cline functions defined in Ex 1.45 and the Plus Rule for Cline Functions to define the cline functions of *F*:(*P*1 *ω*)+*P*3 *ω* and *G*:(*P*0 *ω*)+(*P*2 *ω*)+*P*5 *ω*. The cline *CF* should be defined in terms of *CP*1 and *CP*3, while *CG* should be defined in terms of *CP*0, *CP*2, and *CP*5.

1.48 Use the cline functions defined in Ex 1.45 and the Times Rule for Cline Functions to define the cline functions of *F*:(*P*2 *ω*)×*P*3 *ω* and *G*:(*P*0 *ω*)×*P*4 *ω*. Test your definitions by using the fact that $\underline{C}F$ must be equivalent to $\underline{C}P5$ and $\underline{C}G$ must be equivalent to $\underline{C}P4$.

1.49 Suppose that a square expands and contracts in such a way that at time *T* the length of its sides is *L T*, and suppose that $\underline{C}L$ is the cline function of *L*. Use the Times Rule for Cline Functions to define the cline function $\underline{C}A$ of the area function *A*:(*L* ω)×*L* ω. The following graph shows a partitioned square whose area is *A T*+*S*; the area of the inside square is *A T*. Each of the remaining three rectangles has area equal to *S* times the value of one of the terms in the definition of $\underline{C}A$. Correlate the rectangles with the terms.

```
 --------------        ---
|             | |       ↑
|-------------|-|       |
|             | |       |
|             | |       |
|     A T     | |    L T+S
|             | |       |
|             | |       |
|             | |       ↓
 --------------        ---
```

1.50 Show for a scalar *B* that the cline function of *F*:*B*+0×ω is $\underline{C}F$:0×α+ω. Then use the Times Rule for Cline Functions to show that for a function *G* and its cline function $\underline{C}G$, the cline of *H*:*B*×*G* ω is $\underline{C}H$:*B*×α $\underline{C}G$ ω.

1.51 Use the cline functions in Ex 1.45, the Plus Rule for Cline Functions and Ex 1.50 to define the cline functions of each of the following functions.

a. *F*:3×*P*2 ω
b. *G*:(*P*2 ω)-*P*3 ω
c. *H*:(4×*P*1 ω)+(¯3×*P*2 ω)+4×*P*4 ω

1.52 Repeat Ex 1.45 for the reciprocal functions of the functions in that exercise.

1.53 Use the cline functions defined in Ex 1.45 and the Composition Rule for Cline Functions to define the cline functions of *F*:*P*2 *P*3 ω and *G*:*P*4 *P*1 ω. Test your definitions by using the fact that $\underline{C}F$ must be equivalent to the cline of *P*6:ω*6 and $\underline{C}G$ must be equivalent to $\underline{C}P4$.

1.54 Repeat Ex 1.49 using the Composition Rule for Cline Functions and the fact that the area function can be defined by $\underline{A}$:*P*2 *L* ω, where *P*2:ω*2.

1.55 Listed below are definitions of functions *F* and *G*. In each case show that *G* is equivalent to the cline function of *F*.

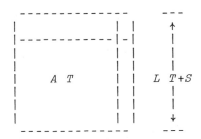

F	*G*
3×ω*2	(3×α)+6×ω
1+ω+ω*2	1+α+2×ω
4+(¯3×ω)+5×ω*2	¯3+(5×α)+10×ω

Using these examples as a guide, define the cline function of *A*+(*B*×ω)+*C*×ω*2 for scalars *A*, *B*, and *C*.

1.56 Use Ex 1.55 to show that if *F*:*A*+(*B*×ω)+*C*×ω*2 then

```
0 C̲F ω
B+2×C×ω
```

Then use this identity to define the derivative of each of the following functions (for example, the derivative of *F*:3+(5×ω)+¯7×ω*2 is *D̲F*:0 *C̲F* ω, which according to the above identity is *D̲F*:5+2×¯7×ω, or equivalently *D̲F*:5+¯14×ω).

a. *F*:4+(¯1×ω)+3×ω*2
b. *G*:2+3×ω*2
c. *H*:5-ω*2
d. *I*:5+6×ω
e. *J*:4×ω*2

1.57 The graphing strategy of Section 1.2 uses pairs of neighboring points on a graph to produce segments of the graph. These segments represent the local behavior of the function near pairs of points. As illustrated in Section 1.5, local behavior can also be represented by one point on the graph together with a segment of the line tangent to the graph at that point. The purpose of this exercise is to study a graphing strategy based on the latter representation of local behavior. The graph of each of the functions defined below is to be sketched over the interval (¯5≤ω)∧5≤ω. The strategy to be used is as follows:

Step A. Choose two integer arguments, for example ¯3 and 4, and plot the two points on the graph of the function for these arguments. Determine the slopes of the lines tangent to the graph at these points and sketch segments of the tangent lines near these points. Then sketch segments of the graph near these points. Try to sketch the entire graph by connecting these segments with a smooth curve.

Step B. Choose a new integer argument, say 2, and plot the point on the graph of the function for this argument. Try to sketch the entire graph using this new information.

Step C. Determine the slope of the line tangent to the graph of the function at the point chosen in step B and sketch a segment of this line near the points. Try to sketch the entire graph using this new information.

Step D. Repeat steps B and C until all the integer arguments between ¯5 and 5 have been used.

The functions to be graphed:

a. ¯12+α+α*2
b. 1+α+α*2
c. 8+(¯2×α)+¯1×α*2

1.58 a. Sketch the graph of the function *F*:*A*+0×ω, where *A* is a scalar. Devise a geometric argument to show that D̲*F*:0×ω. That is, what is the slope of the line tangent to the graph of *F* at a coor dinate point *X*,*F X*?

b. Sketch the graph of the function *F*:*A*+*B*×ω where *A* and *B* are scalars. Devise a geometric argument to show that D̲*F*:*B*+0×ω.

1.59 Determine the derivative of each function in Ex 1.45 and then sketch their graphs using the strategy described in Ex 1.57. The relations among the graphs can be seen by sketching them all in one coordinate system.

1.60 Use the Plus Rule for Derivatives and Ex 1.58 to determine the derivative of each of the following functions.

a. ω+ω*3
b. 5+ω*4
c. (3+4×ω)+ω*5
d. 7+(¯5×ω)+ω*3

1.61 Determine the derivative of each of the following functions and then repeat Ex 1.57 for these functions.

a. ¯6+ω+ω*2
b. ¯5+(3×ω*2)+2×ω*3
c. ¯3-ω*2
d. 4-(3×ω)+ω*3

1.62 Determine the derivative of each of the following functions and then repeat Ex 1.57 for these functions.

a. (3×ω*3)+4×ω*5
b. ¯7+(4×ω)+¯5×ω*7
c. +/(ω*0 1 2)×3 7 ¯4
d. +/(ω*0 2 4)×¯2 1 ¯3
e. +/(ω*ι0)×ι9

1.63 Use the Plus Rule for Derivatives and the identity:

(*F* ω)-*G* ω
(*F* ω)+¯1×*G* ω

to show that the derivative of (*F* ω)-*G* ω is (D̲*F* ω)-D̲*G* ω. This is, in effect, a minus rule for derivatives.

1.64 Determine the derivative of each of the following:

a. (3+ω)×ω+ω*2
b. ω×ω*2
c. (ω-2×ω*3)×ω*2
d. ×/ω*1 2 3
e. ω-(ω+1)×ω+3×ω*2
f. ×/10ρω

1.65 Mathematical induction can be used to show for a positive integer `N` that the derivative of `ω*N` is `N×ω*N-1`:

a. Use the cline function of `ω*1` to show that its derivative is `1×ω*0`;

b. Assume that the initial statement is true for some positive integer, say `K`. That is, assume that the derivative of `F:ω*K` is `D̲F:K×ω*K-1`. Then show that the derivative of `H:ω*K+1` is `D̲H:(K+1)×ω*K` by verifying the following sequences of identities:

```
H ω
ω×F ω
```

and

```
D̲H ω
(1×F ω)+ω×D̲F ω
(ω*K)+ω×K+ω*K-1
(ω*K)+K×ω*K
(K+1)×ω*K
```

1.66 a. Show that the derivatives of `Q2:÷ω*2` and `Q3:÷ω*3` are `D̲Q:-2÷ω*3` and `D̲Q:-3÷ω*4`.

b. More generally, for a negative integer `N` the derivative of `H:ω*N` is `D̲H:N×ω*N-1`. To see that this is true, first define `F:ω*-N`, so that `H ω` is `÷F ω`. Also, since `N` is a negative integer, `-N` is a positive integer and therefore `D̲F:(-N)×ω*(-N)-1`, or equivalently `D̲HF:-N×ω*-N+1`. Now verify the sequence of identities:

```
D̲H ω
-(D̲F ω)÷(F ω)*2
-(-N×ω*-N+1)÷ω*-2×N
N×(ω*-N+1)×ω*2×N
N×ω*N-1
```

1.67 Determine the derivatives of the following functions and repeat Ex 1.57 for these functions.

a. `÷ω`
b. `1+÷ω*2`
c. `ω+÷ω*3`

1.68 Determine the derivative of each of the following:

a. `÷ω+2×ω*2`
b. `÷(ω+1)×ω+2`
c. `÷+/3 4×ω*1 5`

1.69 Use the identity

```
(F ω)÷G ω
(F ω)×÷G ω
```

and the Times and Reciprocal Rules for Derivatives to produce the derivative of each of the following functions:

a. 3÷ω
b. ω÷ω+1
c. +/5 7 ¯1÷ω*1 5 7
d. (1+ω)÷¯2+ω*2

1.70 Use the Composition Rule for Derivatives to determine the derivatives of the following:

a. (3+2×ω*2)*3
b. 1+(1+ω*4)*2
c. +/7 2 9×(+/3 4×ω*1 5)*0 1 2
d. (ω*3)*4

1.71 a. Determine the derivatives of the following functions, and then repeat Ex 1.57 for these functions.

i. ω*2
ii. (ω-2)*2
iii. (ω+4)*2
iv. (3×ω)*2
v. (0.5×ω)*2
vi. (4+2×ω)*2

b. Using part a as a guide, describe the relation between the graphs of the functions *F* ω and *F A*+*B*×ω for scalars *A* and *B*. The function *A*+*B*×ω is called a scale function.

1.72 Sketch the graph of the function 8+(¯6×ω)+(¯3×ω*2)+ω*3 on the interval 3≥|ω. Using the example in the text as a guide, indicate the rate of increase or the rate of decrease near the arguments ¯1.5 and 2.

1.73 Repeat Ex 1.72 for the function 3+(¯3×ω*2)+ω*3.

1.74 Determine an integral of each of the following:

a. 4×ω*6
b. (3×ω)+¯4×ω*2
c. +/9 5 ¯6×ω*0 1 2
d. (ω+3)*4
e. (ω+3)*4
f. +/1 ¯2×(ω-4)*1 4
g. (1+2×ω)*5

1.75 Produce the function *F* for which *DF*:ω+3×ω*2 and 3=*F* 1.

1.76 Repeat Ex 1.75 for each of the following.

a. *DF*:(¯1+3×ω)*2 and 2=*F* 3
b. *DF*:3÷ω*3 and 0=*F* 1
c. *DF*:+/5 7×ω*0 1 and ¯7=*F* 0

1.77 A ball is thrown straight upward from a height of 100 feet with an initial velocity of 20 feet per second.

a. Define the velocity function *VEL* and the position function *ALT* of the ball.

If at time *T*, 1=×*VEL T*, then the ball is still rising, while if ¯1=×*VEL T* the ball is falling. If 0=*VEL T*, then *T* represents the instant at which the ball has stopped rising and will begin to fall. Evidently, at this instant the ball has achieved its maximum altitude.

b. Determine the maximum altitude which the ball attains.

c. How many seconds after the ball is thrown does it hit the ground?

1.78 Repeat Ex 1.77 for an initial height of 50 feet and an initial velocity of 30 feet per second.

1.79 Repeat Ex 1.77 for an initial height of 75 feet and an initial velocity of ¯20 feet per second.

In the next two exercises we will develop an accurate description of the lunar landing problem, allowing for the firing of a retro rocket in bursts of one second duration. We begin by producing an accurate description for each one second interval during which the acceleration of the vehicle is constant. The data used is from Ex 1.29.

1.80 As in Ex 1.29 a firing strategy for the retro rocket is represented by a logical vector *L* for which *ACC L* is the vector of accelerations, where *ACC*:¯6+50×ω; a logical vector is one whose elements are either 0 or 1. For an integer *I* and the time interval (*I*≤ω)∧ω≤*I*+1, the acceleration during this interval is either ¯6 feet per second per second or 44 feet per second per second, depending on whether or not the retro rocket is off or on. In either case the acceleration is constant and therefore the acceleration function during this one second interval can be represented by the scalar *ACC L*[*I*], which has the value ¯6 or 44. Show that the velocity function *VEL* of the landing vehicle during this interval is

VEL:*VI*+(*ACC L*[*I*])×ω-*I*

for a scalar *VI*. That is, show that the derivative of *VEL* is D̲*VEL*:*ACC L*[*I*]. Next show that *VI*=*VEL I*, so that the scalar *VI* must represent the velocity of the landing vehicle at the beginning of the time interval.

Now show that the altitude function *ALT* of the landing vehicle during this interval is

ALT:*AI*+(*VI*×ω-*I*)+(0.5×*ACC L*[*I*])×(ω-*I*)⋆2

for a scalar *AI*. Then show that *AI*=*ALT I*, so that the scalar *AI* must represent the altitude of the landing vehicle at the beginning of the time interval.

Note the important identities for the velocity and altitude of the landing vehicle at the end of this one second interval:

```
VEL I+1                                                  [1]
VI+ACC L[I]
```

and

```
ALT I+1                                                  [2]
AI+VI+0.5×ACC L[I]
```

In words, the velocity at the end of the interval is identical to the velocity at the beginning of the interval plus the constant acceleration during the interval. Also, the altitude at the end of the interval is identical to the altitude at the beginning of the interval plus the velocity at the beginning of the interval plus half of the constant acceleration during the interval.

Suppose that the vector *V* represents the vector of velocities of the vehicle at the junctures of two one second time intervals. That is, the element *V*[*I*] represents the velocity of the vehicle at time *I*, which is also the velocity at the end of the (*I*-1)th time interval as well as the velocity at the beginning of the *I*th time interval. Then according to Identity 1 above:

```
V[I+1]
V[I]+ACC L[I]
```

Consequently, according to the definition of the function *SUM*, if the scalar *V*0 represents the velocity of the vehicle at the beginning of the firing sequence:

```
V                                                        [3]
V0+SUM ACC L
```

Now suppose that the vector *A* represents the vector of altitudes of the vehicle at one second intervals. That is, the element *A*[*I*] represents the altitude at time *I*, which is also the altitude at the end of the (*I*-1)th time interval as well as the altitude at the beginning of the *I*th time interval. Then according to Identity 2 above:

```
A[I+1]
A[I]+V[I]+0.5×ACC L[I]
```

Therefore, if the scalar *A*0 represents the altitude of the vehicle at the beginning of the firing sequence:

```
A                                                        [4]
A0+SUM (¯1↓V)+0.5×ACC L
```

We can compare this accurate description of the lunar landing problem with the approximate description in Ex 1.29. In the second example of that exercise the altitude and velocity at the beginning of the firing sequence are 10000 feet and ¯750 feet per second respectively. The firing sequence was represented by 15↑10ρ1, and the resulting vector of accelerations was

```
      ACC 15↑10ρ1
44 44 44 44 44 44 44 44 44 44 ¯6 ¯6 ¯6 ¯6 ¯6
```

Note that the vector of velocities *V*:

```
      V←¯750+SUM ACC 15↑10ρ1
```

is both the one produced in Ex 1.29 and the one described by Identity 3 above. The vector of altitudes now named *ALT*1 and produced in Ex 1.29 is

```
      ALT1←10000 SUM V
```

The vector of altitudes *ALT*2 described as *A* in Identity 4 above is

```
      ALT2←10000+SUM (¯1↓V)+0.5×ACC 15↑10ρ1
```

¯1↓*ALT*1	*ALT*2
10000	10000
9250	9272
8544	8588
7882	7948
7264	7352
6690	6800
6160	6292
5674	5828
5232	5408
4834	5032
4480	4700
4170	4387
3854	4068
3532	3743
3204	3412
2870	3075

1.81 The computational scheme devised in Ex 1.30 for experimenting with the lunar landing problem carries over to the more accurate description of this problem. The original form of the table remains the same, as does the computation of new elements of the vector of velocities *V*. The only change is in the computation of new elements in the vector of altitudes *A*. In Ex 1.30 a new element *A*[*I*+1] was computed by way of the equation *A*[*I*+1] = *A*[*I*]+*V*[*I*]. In the accurate description of the lunar landing problem, a new element *A*[*I*+1] is to be computed by way of the equation *A*[*I*+1]=*A*[*I*]+*V*[*I*]+0.5×*ACC*[*I*] (see Ex 1.80, Identity 2). For example, in place of the first table in Ex 1.30 we have:

```
 AC                        ¯6        44        ¯6        44       ¯6
  V              ¯250
  A     1500     1247
```

and in place of the completed table in those exercises we have:

```
 AC                        ¯6        44        ¯6        44       ¯6
  V              ¯250     ¯256      ¯212      ¯218      ¯174     ¯180
  A     1500     1247     1013       798       602       425
```

Use this new computational scheme to experiment with the lunar landing problem.

2

2.1 Deterine the matrix product *A*+.×*B* for the following matrices *A* and *B*.

```
        A                          B
¯4  2 ¯1  0               ¯4 ¯4  1
¯3 ¯5  1  1                3 ¯4  0
 4 ¯2  0  3               ¯4  0  3
                           2  6  5
```

2.2 Repeat Ex 2.1 for the matrices *A* and *B* below.

```
        A                          B
¯2  1  2  4                3  2 ¯5
¯2 ¯3  4  2                1  4 ¯3
                          ¯1  3  0
                          ¯3 ¯2 ¯2
```

2.3 The +.× inner product *A*+.×*V* of a matrix *A* and a vector *V* is the vector defined by the identity

```
(A+.×V)[I]
+/A[I;]×V
```

a. Show that *A*+.×*V* equals 52 26 62 54 46, where *A* and *V* are defined as follows:

```
     A                    V
¯1  9 ¯2             0 8 10
 4  2  1
 8  4  3
 9 ¯2  7
 7  7 ¯1
```

b. Show that Identity 2.1.1 is valid for the matrix *A* and the vector *V* substituted for *B*.

2.4 Evaluate `A+.×V` for the following *A* and *V*:

```
      A                      V
4  5  5  8            2  2  4  1
4 ¯1  0  1
```

2.5 The product `W+.×B` of a vector *W* and a matrix *B* is the vector defined by the identity

```
(W+.×B)[I]
+/W×B[;I]
```

a. Show that `W+.×B` equals `8 ¯7 11 ¯3`, where *W* and *B* are defined as follows:

```
      B                W
2 ¯1  2  0          4 ¯3
0  1 ¯1  1
```

b. Show that the Identity 2.1.1 is valid for this matrix *B* and the vector *W* substituted for *A*.

2.6 Evaluate `W+.×B` for the following *B* and *W*:

```
      B                      W
6  3  6 ¯2 ¯1          ¯3 ¯2 1
4 ¯2 ¯2  0 ¯3
1 ¯3  4  7 ¯1
```

2.7 The matrix product is associative. That is, for matrices *A*, *B* and *C*:

```
A+.×(B+.×C)
(A+.×B)+.×C
```

Test this identity.

2.8 The matrix product is not commutative. That is, the identity

```
A+.×B
B+.×A
```

is not valid for all matrices *A* and *B*. Test this identity for the matrices *A* and *B* in Ex 2.1.

2.9 Show that `+.×` distributes over `+`, that is, for matrices *A*, *B* and *C*:

```
A+.×B+C
(A+.×B)+A+.×C
```

by verifying the sequence of identities:

```
(A+.×B+C)[I;J]
+/A[I;]×B[;J]+C[;J]
(+/A[I;]×B[;J])++/A[I;]×C[;J]
(A+.×B)[I;J]+(A+.×C)[I;J]
((A+.×B)+A+.×C)[I;J]
```

2.10 Show that for a matrix *A* and vectors *X* and *Y*:

```
A+.×X+Y
(A+.×X)+A+.×Y
```

2.11 Show that for matrices *A*, *B* and *C*:

```
(A+B)+.×C
(A+.×C)+B+.×C
```

2.12 Show that for vectors *U* and *V* and matrices *C*:

```
(U+V)+.×C
(U+.×C)+V+.×C
```

2.13 The product *V*+.×*W* of vectors *V* and *W* is the scalar defined by the identity

```
V+.×W
+/V×W
```

Show that the +.× inner product of vectors *V* and *W* is commutative. That is:

```
V+.×W
W+.×V
```

2.14 Table 2.2.1 was produced using the fact that if the vector *P* represents the proportions of the market held by the three companies in any given year, then the vector

```
PNEXT←TR P
```

represents the proportions for the following year. Therefore, starting with the proportions for 1970, the proportions for 1971 can be computed, then those for 1972, then 1973 and so on. Reproduce the values in Table 2.2.1.

2.15 If in the example of Section 2.2 we wished to compute the proportions for the second year following a given year, we could apply the computational scheme described in Ex 2.14 twice. Precisely:

```
PNEXT←TR P
PNEXTNEXT←TR PNEXT
```

Let *T*2 denote the matrix defined by the expression *T*+.×*T*. Use the associativity of the matrix product to show that ∧/*PNEXTNEXT*=*T*2+.×*P*, or equivalently ∧/*PNEXTNEXT*=*TR*2 *P*, where *TR*2:*T*2+.×ω. Consequently the vector *PNEXTNEXT* can be produced from the vector *P* by applying the computational scheme described in Ex 2.14 only once if the function *TR* is replaced by the function *TR*2.

2.16 In reference to the example of Section 2.2, define the function *TR*5 such that if the vector *P* represents the percentages of the market held by the three companies in a given year, then the vector *TR*5 *P* represents the percentages five years later.

2.17 Use the function *TR*5 defined in Ex 2.16 to reproduce the values in Table 2.2.1.

2.18 In reference to the example of Section 2.2, test the fact that the proportions of the market held by the three companies will eventually approach the equilibrium distribution 0.28 0.40 0.32 no matter what the initial 1970 proportions may be. For example, perform this test for each of the following 1970 proportions vectors.

```
P70←0.7  0.1   0.2
P70←0.3  0.55  0.15
P70←0.3  0.25  0.45
```

2.19 Write out the interpretations of each element of the matrix *T* in the example of Section 2.2 in terms of customers lost and customers retained by each company. For example, the element *T*[0;0] is the proportion of *A*'s customers that *A* retains. Use these interpretations to explain the fact that ∧/1=+/̸*T*.

2.20 Use the forward substitution algorithm to produce the solution of the equation ∧/*Y*=*A*+.×ω for the following matrix *A* and vector *Y*.

```
     A                          Y
¯4   0   0               ¯16 ¯4 131
 2  ¯2   0
 5   5   9
```

2.21 Repeat Ex 2.20 for

```
     A                          Y
 2   0   0   0           18 ¯90 144 27
¯5  ¯5   0   0
 5   3   8   0
 2  ¯4   4   1
```

2.22 If *A* is a square lower triangular matrix that is not complete, the equation ∧/*Y*=*A*+.×ω may not have a solution.

Use the forward substitution algorithm to show that this equation does not have a solution for the following matrix *A* and vector *Y*.

```
        A                       Y
¯6  0  0  0           0 2 10 70
 4  2  0  0
 1 ¯5  0  0
 4  7  5 ¯3
```

2.23 If *A* is a square lower triangular matrix that is not complete, the equation ∧/*Y*=*A*+.×ω may have more than one solution. Use the forward substitution algorithm to show that this equation has more than one solution for the following matrix *A* and vector *Y*.

```
        A                       Y
 5  0  0  0           20 52 ¯5 6
 8  4  0  0
¯5  3  0  0
 0  5  1 ¯3
```

2.24 Use the backward substitution algorithm to produce the solution of the equation ∧/*W*=*C*+.×ω for the following matrix *C* and vector *W*.

```
      C                     W
8 ¯1  1             28 ¯24 ¯24
0 ¯6 ¯6
0  0 ¯6
```

2.25 Repeat Ex 2.24 for

```
      C                     W
4  2  7  6          125 ¯55 16 9
0 ¯5  3  0
0  0  5 ¯3
0  0  0  3
```

2.26 Using the definition of the function called *SF* as a guide, define the function called *SB* which produces the solution of the equation ∧/*Y*=*A*+.×ω for a complete upper triangular matrix *A* and a vector *Y* by way of the backward substitution algorithm. Use Exs 2.24 and 2.25 to test your definition.

2.27 If *A* is a square matrix and *X* is a vector such that (∨/0≠*X*)∧∧/0=*A*+.×*X*, then *A* is not a valid right argument of the function domino. The reason for this is that both *X* and the vector 0×*X* would be solutions of the equation ∧/0=*A*+.×ω. However, according to the definition in Section 2.3, if *A* is a valid right argument of domino then this equation could have only one solution. For example, the matrix *A* below is not a valid right argument since `A+.×1 ¯3 3 ¯1` is `0 0 0 0`:

```
      A
 0  1  2  3
 4  5  6  7
 8  9 10 11
12 13 14 15
```

Show that each matrix *A* below is not a valid right argument of domino by producing a vector *X* such that `(∨/0≠X)∧∧/0=A+.×X`.

```
      A
0  0  2  2
1  1  7 ¯2
2  2 ¯1  0
3  3  3 ¯2
```

```
      A
0  4  0 12
1  5  3 15
2  6  6 18
3  7  9 21
```

```
      A
2 0 7
5 0 3
1 0 4
```

`A←(N,N)ριN*2` for `N≥3`. Hint. See the above example.

2.28 It is also true that a square matrix *A* is not a valid right argument of domino if there is a vector *W* for which `(∨/0≠W)∧∧/0=W+.×A`. This fact is sometimes useful, for it may be easier to determine such a vector *W* than to determine a vector *X* such that `(∨/0≠X)∧∧/0=A+.×X`. For example, find such vectors *W* and *X* for the matrix

```
      A
4  4  6 ¯1
2  1 ¯1 ¯3
4  4  6 ¯1
9  4  1 10
```

2.29 Exercises 2.29 through 2.32 deal with the application discussed in Section 2.2. Recall the transition matrix

```
      T
0.8              0.1              0.05
0.1              0.85             0.1
0.1              0.05             0.85
```

and the equilibrium distribution vector

```
      E
0.28  0.4  0.32
```

The vector *E* is a solution of the equation `∧/ω=T+.×ω`. Define the matrix *I* as follows:

```
      ⎕←I←(ι3)∘.=ι3
1 0 0
0 1 0
0 0 1
```

Show that `I+.×E` is *E*, and then show that `∧/0=(T-I)+.×E`.

Then show that *T-I* is not a valid right argument of the function domino.

2.30 Define the matrix *R* as follows:

```
      ⎕←R←(T-I)[0 1;],[0] 1 1 1
¯0.2            0.1           0.05
 0.1           ¯0.15          0.1
 1              1             1
```

(*A*,[0] *V* appends *V* to the bottom of *A*.) Use the fact that 1=+/*E* to show that *E* is a solution of the equation

```
      0 0 1=R+.×E
```

The matrix *R* is a valid right argument of the function domino. Consequently, *E* is 0 0 1⌹*R*, which suggests a direct method for producing the equilibrium distribution for any transition matrix *T*. Namely, the equilibrium distribution for the matrix *T* above can be produced as follows:

```
      0 0 1 ⌹R
0.28 0.4 0.32
```

2.31 Use the procedure illustrated in Ex 2.30 to produce the equilibrium distribution vector *E* for the transition matrix

```
      T
0.75       0.05       0.1        0.1
0.1        0.85       0.05       0.1
0.05       0.05       0.8        0.1
0.1        0.05       0.05       0.7
```

2.32 Which of the expressions below, as functions of complete upper triangular matrices *A* and conformable vectors *Y*, is equivalent to *Y SB A*?

a. *Y SF* ⍉*A*　　　　c. (⊖*Y*) *SF* ⍉*A*
b. (⊖*Y*) *SF* ⊖⌽*A*　　d. (⌽*Y*) *SF* ⌽⊖*A*

2.33 The function *L*:*B*+*M*×⍵ is often called a linear function because for scalars *B* and *M*, the graph of this function is a straight line. However, if *B*≠0 this function does not distribute over +, and so is not a linear function according to the definition in Section 2.4. Show that *L* distributes over + if *B*=0 but does not if *B*≠0.

2.34 The monadic function *F* is called an <u>affine function</u> if it distributes over the dyadic function *AV*:0.5×⍺+⍵:

```
      F X AV Y
      (F X) AV F Y
```

Show that the function *L* defined in Ex 2.33 is an affine function by verifying the sequence of identities:

```
L X AV Y                 0.5×(B+M×X)+B+M×Y
B+M×0.5×X+Y              (L X) AV L Y
0.5×B+B+M×X+Y
```

2.35 Show that the function *SUM*:0,+\α of Section 1.1 is a linear function.

2.36 Apply the function *E̲DIFF* to the identity matrix *ID* 4 to produce the result given in Section 2.4. Organize the work by entering the elements of the resulting matrix in the squares of a sheet of graph paper, producing the result one column at a time.

2.37 Since the function *SUM* is a linear function it can be represented as a +.× inner product. Using the definition of the function *E̲DIFF* in Section 2.4 as a guide, define the analogous function *E̲SUM*, using *SUM* in place of *DIFF*. Then use the procedure suggested in Ex 2.36 to produce the matrix *M*←*E̲SUM ID* 4. Test the identity

```
SUM V
M+.×V
```

for the vectors *V*←1 2 5 6 and *V*←0 ¯4 2 1.

2.38 The function ⊖ω distributes over every dyadic scalar function. In particular, this function distributes over + and so is a linear function. Use the procedure suggested in Ex 2.36 to produce the matrix *M* such that for vectors *V* with 5=ρ*V*:

```
⊖V
M+.×V
```

Test this identity for the vectors *V*←2 4 ¯6 8 3 and *V*←1 0 ¯4 5 6.

2.39 Repeat Ex 2.38 with the function 1 0 1 1 1/ω in place of ⊖ω.

2.40 Repeat Ex 2.38 with the function 2↑ω.

2.41 Repeat Ex 2.38 with the function 3↓ω.

2.42 To display matrices of the form *E̲F ID N* on graph paper we apply *F* column-by-column to the identity matrix *ID N*. Analogously, to display matrices of the form ⍉*E̲F* ⍉*ID N* we should apply *F* row-by-row to the matrix *ID N*. Produce the matrix *O* defined by O←⍉ESUM ⍉ID 5. Tezt the identity

```
SUM V
V+.×O
```

for the vectors V←¯5 3 2 0 4 and V←1 0 7 6 ¯2.

2.43 Determine the matrix *O* such that for vectors *V* for which 5=ρV:

```
⌽V
V+.×O
```

Test this identity for the vectors V←2 ¯5 0 4 7 and V←1 0 ¯3 2 6.

2.44 Repeat Ex 2.43 for the function 1 0 1 1 0 1 1\ω in place of ⌽ω.

2.45 Let M←ESUM ID 5 and O←⍉ESUM ⍉ID 5. Which of the following identities are valid for 5 by 5 matrices *X*?

```
SUM X              SUM X
M+.×X              X+.×O
```

2.46 Repeat Ex 2.27 using the function *DET*. That is, in each case show that 0=DET A.

2.47 Use *DET* to show that the matrix *R* of Ex 2.30 is a valid right argument of domino. That is, show that 0≠DET R.

2.48 Show that if ∧/2=ρA then

```
DET A                    DET A
-/A[;0]]×⌽A[;1]          -/A[0;]×⌽A[1;]
```

2.49 Plot the coordinate points and directed line segments in a plane represented by each of the following nine 2-element vectors:

```
 2  7     ¯3  1     5  0
 4 ¯9      0 ¯3    ¯8  4
¯2  0     ¯1 ¯6     0  7
```

2.50 Determine the point *M* which is the midpoint of the line segment joining the points *V* and *W* for each pair of vectors *V* and *W* below. Plot the points *V*, *W* and *M* in each case.

```
a.  V←3 7; W←7 3          c.  V←6 2; W←¯6 ¯2
b.  V←2 ¯5; W←¯4 1        d.  W←¯5 ¯3; V←¯6 7
```

2.51 Plot the points represented by *W* and *S*×*W* for each vector *W* and scalar *S* below.

a. *W*←7 3; *S*←2; *S*←0; *S*←¯1
b. *W*←¯5 1; *S*←¯3; *S*←1; *S*←0.1
c. *W*←¯6 ¯7; *S*←0.5; *S*←2; *S*←¯2

2.52 For each vector *V* and scalar *D* below, determine a vector *W* which represents a point that is at a distance *D* from the origin and lies on the line passing through the origin and the point represented by *V*. Is there more than one such vector in each case?

a. *V*←3 4; *D*←1; *D*←4; *D*←0.5
b. *V*←¯5 12; *D*←10; *D*←0.1; *D*←1
c. *V*←3 ¯4; *D*←5; *D*←0.5

2.53 Without using recursion define the function *F* which is equivalent to *DIRECTION* for vector arguments and otherwise applies along the last axis.

2.54 a. For vectors *U*, *V*, and *W* form the matrices *M*←3 2⍴*U*,*V*,*W* and *O*←*M*,1 1 1. Then 0.5×|*DET O* is the area of the triangle whose vertices are the points represented by *U*,*V* and *W*. Test this fact for each choice of *U*,*V* and *W* below.

i. *U*←0 0; *V*←0 5; *W*←4 0
ii. *U*←2 0; *V*←3 0; *W*←0 0
iii. *U*←0 0; *V*←1 2; *W*←2 0

b. The sign of *DET O* is related to the order in which the vertices appear in the matrix *M*. For vectors *U*,*V* and *W* below sketch the triangle whose vertices are represented by these vectors, and draw arrows on the sides of the triangle which point from *U* to *V*, from *V* to *W*, and from *W* to *U*. The result is a directed path along the triangle which points either clockwise or counterclockwise. For example, in part a above the path is clockwise, and in part b it is counterclockwise. Make a table indicating clockwise or counterclockwise for each choice of *U*, *V* and *W* below, and also showing ×*DET O*. Then state the relation between ×*DET O* and the order of the points.

iv. *U*←3 2; *V*←4 7; *W*←0 1
v. *U*←¯3 1; *V*←0 ¯3; *W*←2 2
vi. *U*←3 2; *V*←0 1; *W*←4 7
vii. *U*←0 ¯3; *V*←¯3 1; *W*←2 2

3

3.1 The polynomials with coefficient vectors *C* and *D* are said to be <u>equivalent</u> if

```
+/C×ω*ιρC
+/D×ω×ιρD
```

Show that the polynomials with coefficient vectors `2 3 0 ¯5 0 0` and `2 3 0 ¯5` are equivalent by verifying the sequence of identities

```
+/2 3 0 ¯5 0 0 ×ω*ι6
(+/2 3 0 ¯5 ×ω*ι4)++/0 0×ω*4 5
+/2 3 0 ¯5×ω*ι4
```

3.2 For a non-negative integer *N* and a vector *C* show that the polynomials with coefficients vectors *C* and `(N+ρC)↑C` are equivalent.

The last two elements of the coefficient vector `2 3 0 ¯5 0 0` and, in general, the last *N* elements of a coefficient vector of the form `(N+ρC)↑C` are called <u>trailing zeros</u>. Note that a coefficient vector with no <u>trailing zeros</u> is of normal form.

3.3 A vector *C* is in normal form if and only if `0≠¯1↑C`. Test the fact that for a vector *C*, the vector defined by the expression `(1+DEGREE C)↑C` is in normal form and that the polynomials with coefficient vectors *C* and `(1+DEGREE C)↑C` are equivalent.

3.4 Determine the degree and order of each of the following:

a. `1 ¯7 3` c. `12↑1 0`
b. `0 2 0 1 0 0` d. `¯9↑¯4 0 1 0`

3.5 For a scalar *X* show that the function `+/α×X*ιρα` is a linear function.

3.6 There is a convenient scheme for producing the coefficient vector of the derivative of a polynomial. For example, the vector `DERIV 3 2 0 ¯2 0 1` can be produced as follows:

```
   3 2 0 ¯2 0 1
 × 0 1 2  3 4 5
 ---------------
     2 0  6 0 5
```

```
      DERIV 3 2 0 ¯2 0 1
2 0 ¯6 0 5
```

Use this scheme to produce the vector *DERIV C* for each vector *C* below.

```
      a. C←0 1 0 4 3.12 ¯9
      b. C←4 2 0 0 7 ¯8 0
      c. C←⍳20
```

3.7 Test the fact that for vectors *C*:

```
      DEGREE DERIV C
      ¯1+DEGREE C
```

For example, test this identity for each of the following vectors.

```
      a.  C←1 3 ¯2 1
      b.  C←1 0 ¯4 0 ¯5 0 0
      c.  C←0 0 5 0 ¯7 0
```

3.8 Test the fact that the function *DERIV* is a linear function. For each pair of vectors *A* and *B* below, use the procedure described in Ex 3.6 to produce the vectors *DERIV A* and *DERIV B* and *DERIV A+B*. Then compare the vectors *DERIV A+B* and (*DERIV A*)+*DERIV B*.

```
      a.  A←3 ¯4 2; B←0 2 7
      b.  A←4 0 0 1; B←2 ¯1 0 5
      c.  A←9 ¯9 5 2 1; B←0 6 5 ¯1 3
```

3.9 Show that column *I* of an identity matrix *ID N* is a coefficient vector representing the polynomial *ω⋆I*. Formally:

```
      (ID N)[;I] POLY ω
      ω⋆I
```

3.10 Use the computation scheme of Ex 3.6 to test the identity

```
      DERIV (ID N)[;I]
      I×1↓(ID N)[;I]
```

for each of the following pairs of scalars *N* and *I*:

```
      a. N←5; I←3
      b. N←4; I←1
      c. N←6; I←0
```

3.11 It follows from Ex 3.8 that the function *DERIV* can be represented as a +.× inner product. Using Ex 3.10, it is easy to display a matrix involved in this representation. For example, the matrix defined by the expression

EDERIV ID 4 can be displayed as follows:

```
0 1 0 0
0 0 2 0
0 0 0 3
```

Display the matrices defined by the expression *EDERIV ID K* for *K*←5; *K*←7; *K*←8.

3.12 For a vector *C* the polynomial with coefficient vector *DERIV DERIV C* is called the <u>second derivative</u> of the polynomial with coefficient vector *C*, the polynomial with coefficient vector *DERIV DERIV DERIV C* the <u>third derivative</u>, and so on. (In this regard the polynomial with coefficient vector *DERIV C* is called the <u>first derivative</u> of the polynomial with coefficient vector *C*.)

The vector *DERIV DERIV C* can be produced by a scheme similiar to the one used for *DERIV C*. For example, for the vector 2 7 0 3 0 1:

```
   2 7 0  3 0  1
   0 1 2  3 4  5
×    0 1  2 3  4
   --------------
       0 18 0 20
```

Produce each of the following vectors.

```
DERIV DERIV 3 ¯7 2 0 0 1
DERIV DERIV 10 11 9 0 4 0 2
DERIV DERIV DERIV 4 0 3 ¯7 2
```

3.13 Determine the first, second, and third derivatives of each of the following functions.

```
a.  +/3 0 2 2 ¯1×ω*⍳5
b.  +/¯2 5 7 ¯3 9 1×α*⍳6
c.  +/3 1 2×ω*⍳3
```

3.14 Show that if *F* and *G* are linear functions then the function *F G ω* is also a linear function by verifying the sequence of identities:

```
F G X+Y
F (G X)+G Y
(F G X)+F G Y
```

3.15 It follows from Exs 3.8 and 3.14 that the function *PD2:DERIV DERIV ω* is a linear function. Display on graph paper the matrix defined by the expression *EPD2 ID K* for *K*←4; *K*←6; *K*←9.

3.16 Each vector *D* below is defined in terms of the function *DERIV* and a vector *C*. In each case determine a matrix *M* such that *D*=*M*+.×*C* and test the equation ∧/*D*=*M*+.×*C*:

```
a.  D←DERIV C←3 9 15 0 2
b.  D←DERIV C←0 ¯1 0 0 1 0
c.  D←DERIV DERIV C←1 ¯2 3 ¯4 6 1
```

3.17 As with the function *DERIV*, there is a convenient scheme for producing values of the function *INTEG*. For example, the value `2 INTEG 1 ¯3 6 2` can be produced as follows:

```
     1   ¯3 6   2
  ÷  1    2 3   4
  ----------------
  2  1  1.5 2 0.5
```

Use this scheme to produce each of the following vectors.

```
a.  1 INTEG 1 0 3 1 ¯2
b.  ¯2 INTEG 0 2 7.5 4
c.  3 INTEG 2 5 0 ¯8 1 2
```

3.18 For each vector *C* below determine the vector *IC* for which `∧/C=DERIV IC` and the accompanying proposition is true.

```
a.  C←0 1 3 1; 2=IC POLY 0
b.  C←1 3 1; ¯4=IC POLY 3
c.  C←¯4 2 0 3; 5=IC POLY ¯2
```

3.19 Test the fact that `0 INTEG ω` is a linear function, while, in general, `S INTEG ω` is an affine function (Ex 2.34).

3.20 Determine *M* such that for 4-element vectors *C*:

```
0 INTEG C
M+.×C
```

3.21 Repeat Ex 3.20 for vectors of length 7.

3.22 If `(ρA)≠ρB`, then the vector defined by `(M↑A)+(M←(ρA)⌈ρB)↑B` is the coefficient vector of a polynomial equivalent to the sum of the polynomials with coefficient vectors *A* and *B*. Prove this by verifying the sequence of identities:

```
(+/A×ω*ιρA)++/B×ω*ιρB
(+/(M↑A)×ω*ιM)++/(M↑B)×ω*ιM
+/((M↑A)+M↑B)×ω*ιM
```

Compare the above expression with the definition of the function *PLUS* in the text.

3.23 Show that for vectors *A* and *B* such that (*DEGREE A*)≠*DEGREE B*:

```
DEGREE A PLUS B
(DEGREE A)⌈DEGREE B
```

3.24 Determine *A* and *B* such that (*DEGREE A*)=*DEGREE B* and (*DEGREE A PLUS B*)<*DEGREE A*.

3.25 Show that for scalars *X*,

```
+/0 0 0 4 7 1×X*⍳6
(X*3)×+/4 7 1×X*⍳3
```

3.26 Show that for scalars *T*,

```
+/0 0 ¯5 0 6 0×T*⍳6
(T*2)×+/¯5 0 6×T*⍳3
```

3.27 Show that for a non-negative integer *N* and a coefficient vector *C*:

```
((-N+⍴C)↑C) POLY ⍵
(⍵*N)×C POLY ⍵
```

3.28 For each pair of vectors *A* and *B* below, determine the vector *A TIMES B* by forming the multiplication table *A*∘.×*B* and summing the upper-right to lower-left diagonals.

```
A←1 3; B←¯2 1 4
A←1 0 2 0 ¯1; B←1 2 0 ¯5
```

3.29 Repeat Ex 3.28, reversing the roles of *A* and *B*, and compare the results to conclude that *TIMES* is commutative.

3.30 The function *TIMES* is linear in either argument. Show that *TIMES* is linear in the right argument by verifying the sequence of identities

```
A TIMES B+C
+⌿(-⍳⍴A)⌽A∘.×(B+C),0×1↓A
(+⌿(-⍳⍴A)⌽A∘.×B,0×1↓A)++⌿(-⍳⍴A)⌽A∘.×C,0×1↓A
(A TIMES B) + A TIMES C
```

3.31 Show that

```
A TIMES (ID N)[;I]
(I⍴0),A
```

3.32 It follows from Ex 3.30 that the function *A TIMES* ⍵ can be represented as a +.× inner product. Using Ex 3.31, it is easy to display the matrices involved in this representation. For example, `1 ¯3 7 E̲TIMES ID 5` yields:

```
 1   0   0   0   0
¯3   1   0   0   0
 7  ¯3   1   0   0
 0   7  ¯3   1   0
 0   0   7  ¯3   1
 0   0   0   7  ¯3
 0   0   0   0   7
```

Display the matrix defined by the expression `A E̲TIMES ID K` for each `K` and `A` below:

```
a. K←5; A← ¯2 0  3 1
b. K←3; A← 1 3 ¯2 4 ¯6 3
c. K←1; A← 1 0  2 3
d. K←4; A← ¯2 1
e. K←4; A← 3 0  1 0 0
```

3.33 For each pair *C* and *D* below determine *M* such that:

```
C TIMES D
M+.×D
```

```
a. C←1 ¯3 3 0; D←4 7 ¯9 0 1
b. C← ¯4 1; D←0 3 0 4 0 1
c. C←3 7 1; D←15 20 0 10
```

3.34 For each case of Ex 3.28 test the identity

```
DEGREE A TIMES B
(DEGREE A)+DEGREE B
```

3.35 Suppose that *A* and *B* and *Q* are vectors such that `∧/B=A TIMES Q`. It follows from the definition of *TIMES* that `(ρB)= ¯1+(ρA)+ρQ`. Assume that there is a vector *R* for which `(ρR)<ρA` and `∧/B=R PLUS A TIMES Q`. Show that

```
(ρQ)=1+(ρB)-ρA
```

3.36 For vectors *A* and *B* there exists a unique pair of vectors *Q* and *R* such that

```
(DEGREE R)<DEGREE A          [1]
∧/B=R PLUS A TIMES           [2]
```

Q is called the <u>quotient</u> of *B* divided by *A* and *R* is called the <u>remainder</u>. Also, the polynomial `Q POLY ω` is called the quotient of `B POLY ω` divided by `A POLY ω` and `R POLY ω` is called the remainder.

Show that if `(DEGREE B)<DEGREE A` then `1ρ0` is the quotient of *B* divided by *A* and *B* is the remainder.

3.37 If *(DEGREE A)≤DEGREE B* then the quotient *Q* and the remainder *R* of *B* divided by *A* can be produced as illustrated in the following example.

```
      A
3 5 ¯2 0 0

      B
9 ¯3 ¯3 46 ¯9 ¯2
```

First, reduce *A* to normal form.

```
      ⎕←A←(1+DEGREE A)↑A
3 5 ¯2
```

The size of the vector *Q* can be determined by *RQ*←1+(ρ*B*)-ρ*A*. Next, form the matrix *N* for which ∧/(*A TIMES Q*)=*N*+.×*Q*:

```
      ⎕←N←A ETIMES ID RQ
 3  0  0  0
 5  3  0  0
¯2  5  3  0
 0 ¯2  5  3
 0  0 ¯2  5
 0  0  0 ¯2
```

The quotient *Q* is produced from the last *RQ* rows of *N* and the last *RQ* elements of *B* as follows:

```
      ⎕←M←((RQ×¯1 1)↑N
 ¯2  5  3  0
  0 ¯2  5  3
  0  0 ¯2  5
  0  0  0 ¯2

      ⎕←Q←((-RQ)↑B)⌹M
2 ¯4 7 1
```

Now we can produce the remainder *R*.

```
      ⎕←R←(DEGREE A)↑B-A TIMES Q
3 ¯1

      B
9 ¯3 ¯3 46 ¯9 ¯2

      R PLUS A TIMES Q
9 ¯3 ¯3 46 ¯9 ¯2
```

Note that the matrix *M* described in the above procedure is a complete upper triangular matrix. Consequently the vector *Q* can be produced by the backward substitutionzalgorithm, in which case the method for producing *Q* is called <u>synthetic division</u>.

Determine the quotient *Q* and the remainder *R* of *B* divided by *A* for each pair of vectors *A* and *B* below.

```
a.  A←2 0 3; B←20 11 27 7 0 ¯3
b.  A←2 3 0 ¯5; B←2 ¯2 ¯19 ¯17 35 ¯5
c.  A←8 3 ¯3 1 0; B←¯41 ¯4 17 ¯7 1 0 0
```

3.38 For a vector *R* the function *F*:×/ω-*R* is equivalent to a polynomial; the elements of *R* are called the <u>roots</u> or <u>zeros</u> of this polynomial since 0=*F R*[*I*] for each index *I*. In this exercise we will define the function *CFR* (for <u>c</u>oefficients <u>f</u>rom <u>r</u>oots) which yields the coefficient vector of an equivalent polynomial. If *R* is a scalar then ×/ω-*R* is ω-*R*, or equivalently ((-*R*),1) *POLY* ω. If *R* is a vector then ×/ω-*R* is (*PTR* (-*R*)∘.*1 0) *POLY* ω. For example:

```
      R←2 ¯4 7

      (-R)∘.*1 0
¯2  4 ¯7
 1  1  1

      C←PTR (-R)∘.*1 0
      C
56 ¯22 ¯5 1

      C POLY R
0 0 0
```

Thus *CFR*:*PTR* (-ω)∘.*1 0. Use *CFR* to produce a coefficient vector of each of the following polynomials.

```
a.  ×/ω-5 ¯3 1
b.  ×/ω-3 0 2 0 ¯1
c.  ×/ω-1 1 1
```

3.39 For the vector ¯1 0 4 we have

```
      CFR ¯1 0 4
0 ¯4 ¯3 1
```

Use this coefficient vector to produce the coefficient vector of the polynomial with roots ¯1 0 4 and whose value at ¯2 is 6.

Exs 3.40-43 are concerned with another definition of *CFR*.

3.40 The vector *P* is called a <u>permutation vector of order</u> ρ*P* if ∧/1=+/*P*∘.=ιρ*P*. In words, the elements of *P* are a rearrangement of the elements of ιρ*P*. For example, 2 1 3 0 4 is a permutation vector of order 5.

If *P* is a permutation vector then the matrix defined by the expression *P*∘.=⍳⍴*P* is called a permutation matrix. For a vector *X* and a permutation vector *P* of order ⍴*X*:

```
X[P]
(P∘.=⍳⍴P)+.×X
```

Test this identity for each pair of vectors *X* and *P* below.

```
X←1 ¯8 3; P←1 2 0
X←2 0 ¯2 1 5; P←2 0 4 1 3
X←1 2 ¯2 4; P←3 2 1 0
```

3.41 The monadic function *F* is called a symmetric function if for every vector of arguments *X* and every permutation vector *P* of order ⍴*X*:

```
F X
F X[P]
```

In words, the values of the function *F* depend only on the elements of an argument vector *X* and not on the order in which the elements appear in *X*. For example, the function +/⍵ is a symmetric function but -/⍵ is not.

Which of the following functions are symmetric functions? For those functions which are not symmetric functions, determine a vector *X* and a permutation vector *P* for which the above identity is not valid.

a. +/⍵
b. ×/⍵
c. */⍵
d. *CFR* ⍵
e. ×\⍵
f. (×/0 1 1/⍵)+(×/1 0 1/⍵)+×/1 1 0/⍵

3.42 For a positive integer *N* the columns of the matrix

```
M←(N⍴2)⊤⍳2*N
```

are the base 2 representations of the integers between 0 and ¯1+2**N* (the primary function ⊤ is called the representation function). For example:

```
      ⎕←M←2 2 2⊤⍳2*3
0 0 0 0 1 1 1 1
0 0 1 1 0 0 1 1
0 1 0 1 0 1 0 1

      +/(⌽M[;3])×2*0 1 2                 +/(⌽M[;5)]×2*0 1 2
3                                  5
```

For a vector *X*, the columns of the matrix

```
M←((ρX)ρ2)⊤⍳2ρX
```

also represent the subsets of the set of elements of *X*. For example:

```
      X
3 ¯9 4

      ⎕←M←((ρX)ρ2⊤⍳2*ρX
0 0 0 0 1 1 1 1
0 0 1 1 0 0 1 1
0 1 0 1 0 1 0 1

      M[;3]/X
¯9 4
```

Element *I* of the vector defined by the expression *X*×.⋆*M* is identical to the product of the elements of the subset of the elements of *X* represented by column *I* of *M*. Continuing the above example:

```
      X×.⋆M
1 4 ¯9 ¯36 3 12 ¯27 ¯108

      (X×.⋆M)[3]                   ×/M[;3]/X
¯36                          ¯36
```

Consider the function

```
ESF:(α∘.=+⌿((ρω)ρ2)⊤⍳2⋆ρω)+.×ω×.⋆((ρω)ρ2)⊤⍳2⋆ρω
```

Argue for the fact that for an integer *I*, the function *I ESF ω* is a symmetric function. This function is called the *I*th <u>elementary symmetric function</u>.

3.43 Test the fact that the values of the function (⌽⍳1+ρω) *ESF* -ω are identical to the values of *CFR ω* for each vector *R* below:

```
a.  R←1 ¯3 2            c.  R←0 2 ¯7 3
b.  R←1 0 5             d.  R←1 0 ¯5 1
```

3.44 Test the fact that for vectors *A* and *B* the matrix defined by *B*∘.+0×1↓α has ¯1+ρ*A* columns, each identical to *B*. Then test the fact that the matrix defined by *PTS B*∘.+0×1↓*A* has ρ*A* columns and column *I* of this matrix is a coefficient vector of the polynomial (*B POLY ω*)⋆*I*. For instance, these tests can be carried out for the following vectors *A* and *B*:

```
A←3 4 ¯1; B←1 0 3 4
A←2 0 ¯7 1; B←2 3 ¯5
```

3.45 The function *PCM*:*ATS* ⍉((¯1+⍴⍺),⍴⍵)⍴⍵ will produce the matrices of Ex 3.44; for this function and *COMP* we have the identity

```
A COMP B
(A PCM B)+.×A
```

Show that

```
(A COMP B) POLY ⍵
A POLY B POLY ⍵
```

by verifying the following sequence of identities:

```
(A COMP B) POLY ⍵
((⍴A COMP B) VAN ⍵)+.×A COMP B
((⍴A COMP B) VAN ⍵)+.×(A PCM B)+.×A
(((⍴A COMP B) VAN ⍵)+.×A PCM B)+.×A
((⍴A) VAN B POLY ⍵)+.×A
A POLY B POLY ⍵
```

3.46 a. Parts b and c of this exercise, through Ex 3.51, are concerned with the effect of applying the function *DIFF* to vectors of values of polynomials. It will be helpful to experiment with the effect before proceding. For each pair of vectors *C* and *X* below evaluate *C POLY X*, *DIFF C POLY X*, *DIFF DIFF C POLY X*, and so on.

```
C←1 ¯3 2; X←6+⍳10
C←2 0 4 1; X←¯9+2×⍳12
```

b. Use the function *PCM* of Ex 3.45 to define the function *APOLY* such that for vectors *C* and scalars *S*:

```
C POLY ⍵+S
((S APOLY C)+.×C) POLY ⍵
```

Hint. The polynomial ⍵+*S* is equivalent to (*S*,1) *POLY* ⍵.

c. Show that the matrix *S APOLY C* is square, upper triangular and has diagonal elements all equal to 1. Also, show that ⍴(*S APOLY C*)+.×*C* equals ⍴*C* and that ¯1↑(*S APOLY C*)+.×*C* equals ¯1↑*C*.

3.47 Use the function *APOLY* of Ex 3.46 to define the function *PD* such that for a vector *C* and scalar *S*:

```
(C POLY ⍵+S)-C POLY ⍵
(S PD C) POLY ⍵
```

Show that (⍴*C*)=⍴*S PD C* and 0=¯1↑*S PD C*; it then follows that *DEGREE S PD C* is at most one less than *DEGREE C*.

3.48 For scalars *A* and *S* let *X←A+S×⍳K* and verify the following sequence of identities:

```
DIFF C POLY X
(1↓C POLY X)-¯1↓C POLY X
(C POLY 1↓X)-C POLY ¯1↓X
(C POLY S+¯1↓X)-C POLY ¯1↓X
(S PD C) POLY ¯1↓X
```

Therefore:

```
DIFF C POLY A+S×⍳K
(S PD C) POLY A+S×⍳K-1
```

The point is that if *DIFF* is applied to a vector of values *C POLY A+S×⍳K*, the result is also a vector of values of a polynomial, and the degree of the latter polynomial is less than *DEGREE C*.

3.49 Define the function *PD2* such that for scalars *A* and *S*:

```
DIFF DIFF C POLY A+S×⍳K
(S PD2 C) POLY A+S×⍳K-2
```

Show that *DEGREE S PD2 C* is at most two less than *DEGREE C*.

3.50 Continue the arguments of Exs 3.48-49 to show that *DEGREE C* applications of the function *DIFF* to the vector of values *C POLY A+S×⍳K* produces a vector that is identical to the vector of values of a polynomial of degree 0, and consequently any two elements of this vector are equal. This fact is sometimes called the <u>Fundamental Theorem of Finite Differences</u>. For example, the degree of the polynomial with coefficient vector 1 ¯3 4 2 is 3, and the third difference of the vector of values 1 ¯3 4 2 *POLY* 6+⍳10 is (10-3)⍴12.

Each vector listed below is of the form *C POLY A+S×⍳K* for a vector *C* and scalars *A* and *S*. Determine the degree of the polynomial in each case.

```
a.  16 29 36 67 92 121 154 191 232
b.  23 1 11 53 127 233 371 541
c.  ¯449 ¯259 ¯131 ¯53 ¯13 1 1 ¯1 37
d.  99 19 3 3 ¯29 ¯141 ¯381
```

3.51 The Fundamental Theorem of Finite Differences of Exercise 3.50 suggests a test indicating, but not proving, whether or not a function is a polynomial. For example, consider the matrix

```
A←3 3⍴2 1 0 ¯3 0 4 1 ¯1 2
```

and the function *F:DET A-⍵×ID 1↑⍴A*. In order to apply this function to vector arguments, consider the extended domain function

```
      EF:(F ω[0]),EF 1↓ω : 0=ρω : ι0

      V←EF ι10
      V
18 10 4 ¯6 ¯26 ¯62 ¯120 ¯206 ¯326 ¯486

      DIFF V
¯8 ¯6 ¯10 ¯20 ¯36 ¯58 ¯86 ¯120 ¯160

      DIFF DIFF V
2 ¯4 ¯10 ¯16 ¯22 ¯28 ¯34 ¯40

      DIFF DIFF DIFF V
¯6 ¯6 ¯6 ¯6 ¯6 ¯6 ¯6
```

The last vector suggests that *F* is a polynomial of degree 3. Checking another vector of values:

```
      DIFF DIFF DIFF EF 3+.1×ι8
¯0.006 ¯0.006 ¯0.006 ¯0.006 ¯0.006
```

This vector also suggests that *F* is a polynomial of degree 3. In fact *F* is a polynomial of degree 3, called the <u>characteristic polynomial</u> of the matrix *A*.

Test the fact that *F* is a polynomial for each matrix *A* below. What is the relation between the degree of the polynomial and ρ*A*?

```
a.  A←2 2ρ3 4 0 ¯7
b.  A←2 0 ¯4∘.×1 8 ¯3
c.  A←¯2 0 1 2∘.*0 1 2 3
```

3.52 For each integer *N* and vector *C* below determine the *N*th order reciprocal of *C* and its remainder.

```
a.  N←8; C←1 ¯1                   c.  N←6; C←1 ¯2
b.  N←8; C←1 ¯1 TIMES 1 ¯1        d.  N←7; C←2 ¯1
```

3.53 Describe the domain of α *REC* ω.

3.54 Show that if the vector *X* is not lean, then the matrix (ρ*X*) *VAN X* has at least two identical rows.

3.55 Show that if a matrix *M* has at least two identical rows, then *M* is not a valid right argument of the function domino. (See Ex 2.27-28).

3.56 Show that if the vector *X* is not lean, then the matrix (ρ*X*) *VAN X* is not a valid right argument of the function domino.

3.57 Consider the monadic function *G*, the vector of arguments *A*+*S*×⍳*K* for scalars *A* and *S*, and the vector of values *Y*←*G A*+*S*×⍳*K*. Recall the test for polynomials described in Ex 3.51. Namely, if *G* is a polynomial of degree *K* and if *DK* denotes the vector resulting from *K* applications of the function *DIFF* to the vector *Y*, then any two elements of the vector *DK* are equal. Resolve the apparent conflict in the following two statements:

a. If the function *G* is not a polynomial, then the test for polynomials, when applied to the vector *Y*, will not indicate that *G* is a polynomial.

b. Even though the function *G* is not a polynomial, the vector *Y* is the vector of values of a polynomial for the vector of arguments *A*+*S*×⍳*K* (namely, the fitting polynomial for the vector of arguments *A*+*S*×⍳*K* and the vector of values *Y*). Consequently the test for polynomials, when applied to *Y*, should not fail.

Hint. When is the test inconclusive?

3.58 One of the important applications of fitting polynomials is to produce the coefficient vector of a polynomial when that vector is not known. For example, consider again the example in Ex 3.51. The function *F* is a polynomial of degree 3 and so has a coefficient vector *C* of length 4. The vector *C* can be produced as follows:

```
      X←1 2 3 4
      Y←E̲F X
      C←X CFP Y

      C POLY ⍳10
18 10 4 ¯6 ¯26 ¯62 ¯120 ¯206 ¯326 ¯486

      E̲F ⍳10
18 10 4 ¯6 ¯26 ¯62 ¯120 ¯206 ¯326 ¯486
```

Use this procedure to produce a coefficient vector for each polynomial in Ex 3.51.

4

4.1 It will be helpful to review the mechanics of differentiation. Use the derivative rules in section 1.5 to produce the derivative of each of the following functions. Then use the graphing strategy described in Ex 1.57 to sketch the graph of each function.

```
a.  ω+3×ω*2
b.  α×1+α*3
c.  ¯7+(8×ω)+ω*2
d.  ÷1+ω
e.  +/(α*⍳5)÷!⍳5
f.  (1-ω*2)÷1+ω*2
g.  ×/3 4 5×ω*0 1 2
h.  ω-7×ω*4
i.  1-α÷1+α*2
j.  +/3 ¯7 0 1×ω*⍳4
k.  +/α*⍳10
l.  ÷/ω*⍳10
```

4.2 For each function *H* below determine functions *F* and *G* for which *H ω* is equivalent to *F G ω*, and then use the composition rule to produce the derivative *D̲H ω*.

```
a.  H:(1-ω*2)*7
b.  H:1 ¯5 1 POLY 1+2×ω
c.  H:0 0 1 POLY 0 0 2 POLY ω
```

4.3 Repeat Ex 4.1 for the following functions.

```
a.  ω*÷3
b.  ω+÷ω*0.5
c.  α×3+α*3.4
d.  3+(4×α*¯2)-α*2
e.  (3+4×ω)*0.5
f.  3 ¯7 0 1 POLY ω*¯1.5
g.  α*⌈/-⍳12
h.  ω*÷/⍳15
```

4.4 Comment on the following statement and then use it to determine the type of the critical argument *A* for each of the functions listed: "If *A*,*F A* is a critical point of *F* then it is a local minimum, an inflection point, or a local maximum according to whether +/(*F A*)>*F A*+*E*,-*E* equals 0,1, or 2, where *E* is a small but non-zero value".

```
A←2  ; 56 144 ¯48 ¯4 3 POLY ω
A←¯3 ; 56 144 ¯48 ¯4 3 POLY ω
A←2  ; 48 ¯18 2 POLY ω
A←4  ; 48 ¯18 2 POLY ω
```

4.5 Another test for the type of a critical argument is as follows. Suppose that *A* is a critical argument of the function *F* and choose a small, non-zero scalar *A*. Form the matrix

```
M←⍉ 2 3⍴(A+E×0 1 ¯1),F A+E×0 1 ¯1
```

What is ×*DET M*,1 1 1 if *A* is a local maximum? Local minimum?
Hint. See Ex 2.54b.

4.6 For each vector *C* below determine the maximum values of the polynomial *C POLY ω* on the interval 3≥|ω.

```
C←1 3 ¯2
C←¯12 9 ¯6 1
C←0 12 ¯6 1
```

Hint. To solve this problem it is necessary to determine the value of each polynomial for only a few arguments, namely for the critical arguments *A* for which *C POLY A* is a local maximum and for the end point arguments 3 and ¯3.

4.7 Repeat Ex 4.6 for minimum values.

4.8 a. What is the largest area among all rectangular regions that can be enclosed by 100 yards of fencing?
Hint. The area is given by ω×50-ω or, equivalently, 0 50 ¯1 *POLY* ω. Determine the maximum value on the interval (0<ω)∧ω<50.

b. A rectangular region next to a building is to be enclosed with 100 yards of fencing; only three sides of the region will have fencing, the side adjacent to the building requiring no fencing. What are the dimensions of the region with the largest area?

4.9 What is the maximum area among all rectangular regions with two vertices on the horizontal axis and the other two vertices above the horizontal axis and on the graph of the polynomial 10 0 ¯1 *POLY* ω?

4.10 Squares of equal size are cut from the corners of a square tin sheet of side length 2 feet and the edges are folded up to form a box. What is the largest possible volume of such a box?
Hint. First define the function *VOL* such that if *X* is the length of a side of the squares cut from the corners, then *VOL X* is the volume of the resulting box.

4.11 A wire 3 feet long is cut in two and one piece is formed into the perimeter of a circle, the other into the perimeter of a square. What is the minimum value of the sum of the areas of the square and circle?

4.12 What is the length of the longest ladder that can be turned through a right angle corner of a six feet wide hallway while remaining parallel to the floor?

4.13 The first derivative test for the type of a critical argument. Suppose that *A* is a critical argument of the function *F*. What is the type of *A* if for scalars *X* and *Y* which are near *A* and such that (*X*<*A*)∧*A*<*Y*:

```
a.  ∧/1 ¯1=×D̲F X,Y
b.  ∧/¯1 1=×D̲F X,Y
c.  0≠+/×D̲F X,Y
```

Repeat Ex 4.4 using the first derivative test.

The next set of exercises deals with the type of problem known as initial value problems and already encountered in Chapter 1, e.g., Ex 1.80.

4.14 a. Define the function *F* whose derivative is *DF*:ω*2 and for which 5=*F* 1.

b. Define the function *F* whose derivative is *DF*:(ω*3)+4×ω*5 and for which 10=*F* ¯2.

It is not always possible to produce a rule of a function which is the solution of an initial value problem as easily as in Ex 4.14. However, this does not mean that we cannot obtain some information about the solution function. Two sources of information are sketches of graphs and tables of approximate values. The following exercises deal with these sources of information for the initial value problems in Ex 4.14.

4.15 In order to sketch the graph of the solution function *F* of the initial value problem in Ex 4.14a (without first defining *F*), first draw the axes and the vertical lines passing through the points *X*,0 for each element *X* of ¯3+0.5×ι13. Consider, for example, the ver- tical line passing through the point 1 0. Since *DF*:ω*2, at the point where the graph of *F* intersects this vertical line the slope of the tangent line to the graph is 1*2, or equivalently 1. Draw small line segments with slope 1 at various locations along this vertical line. Similarly, at the point where the graph of *F* intersects the vertical line passing through 2 0, the line tangent to the graph of *F* has slope 2*2, or equivalently 4. Draw small line segments with slope 4 at various locations along this vertical line. Continue this procedure for each of the vertical lines in the graph. Now, starting at the point 1 5 (Why this point?), sketch the graph of *F* on the interval 3≥|ω. Be sure that your sketch crosses each vertical line with the indicated tangent line slope. Finally, use the rule for *F* produced in Ex 4.14a to sketch the graph of *F* on the interval 3≥|ω and compare the two graphs.

4.16 Repeat Ex 4.15 with Ex 4.14b in place of Ex 4.14a.

4.17 A vector *Y* of approximate values of the solution function *F* of the initial value problem in Ex 4.14a can be produced as follows. Define *X*←¯3+0.5×ι13 and use the procedure described in Ex 1.22 to produce the vector *Y* for which:

```
(DIFF Y)÷DIFF X
(¯1↓X)*2
```

(Why this identity?) and 5=*Y*[8] (Why this element of *Y*?). Plot the points *X*[*I*],*Y*[*I*] for each integer *I* in ι13 and connect each pair of neighboring points with a straight line segment. Compare the resulting <u>polygonal curve</u> with the graphs produced in Ex 4.15.

4.18 Define a rule for the function whose graph is the polygonal curve in Ex 4.17. This function is called the <u>linear interpolation function</u> for the vector of arguments *X* and the vector of values *Y*.

4.19 Repeat Ex 4.17 for the vector *X*←¯3+0.25×⍳25. Which of Exercises 4.17 and 19 gives the best approximations?

4.20 Repeat Ex 4.18 for the polygonal curve in Ex 4.19.

4.21 Repeat Ex 4.17 for the initial value problem in Exercise 4.14b in place of the one in Ex 4.14a. That is, produce the vector *Y* for which:

```
(DIFF Y)÷DIFF X
¯1↓(X*3)+4×X*5
```

and 10=*Y*[2].

4.22 Repeat Ex 4.18 for the polygonal curve in Ex 4.21.

4.23 Repeat Ex 4.21 for the vector of arguments *X*←¯3+0.25×⍳25. Which of Exercises 4.21 and 23 gives the best approximations?

We will now examine two initial value problems whose solutions are not yet known to us. That is, we consider the initial value problems to be implicit definitions of their solutions, and use these implicit definitions to sketch graphs and produce explicit, but approximate, definitions.

4.24 Consider the initial value problem to determine the function *LA* whose derivative is *<u>D</u>LA*:÷⍵ and for which *LA* 1 is 0. (Note that *LA* will be an integral of ÷⍵, or equivalently ⍵*¯1, which is the one function of the form ⍵**N* whose integrals are not yet known to us.) Use the graphing strategy illustrated in Ex 4.15 to produce a sketch of the graph of *LA* on the interval (0≤⍵)∧⍵≤10.

4.25 Repeat Ex 4.17 for the initial value problem in Exercise 4.24 and the vector of arguments 1+⍳10.

4.26 Repeat Ex 4.25 for the vector of arguments 0.5×1+⍳20. Which of Exercises 4.25 and 26 give the best approximations?

4.27 Consider the initial value problem to determine the function *GR* whose derivative is *<u>D</u>GR*:*GR* ⍵ (that is, the derivative of *GR* is identical to *GR*) and for which the value of *GR* at 0 is 2. Devise a variation of the graphing strategy illustrated in Ex 4.15 and produce a sketch of the graph of *GR* on the interval 4≥|⍵.

4.28 Repeat Ex 4.17 for the initial value problem in Exercise 4.27 and the vector of arguments ¯2+0.5×⍳11, using a variation of the procedure in Ex 1.31 to produce the vector of approximate values of *GR*.

4.29 We will now sketch the graph of a function *G* for which:

```
(DG ω)*2
1-(G ω)*2
```

and 1=*G* 0. Note that the functions 1+0×ω and ¯1+0×ω are solutions to this initial value problem. However, there is a more interesting solution and we will assume that the solution *G* is not one of these simple ones.

a. Show that 0 is a local maximum of *G*.

b. Show that as the argument *X* increases from 0, if the values *G X* start to decrease then they will continue to decrease until, if ever, they reach ¯1.

c. Suppose that the values of *G* do reach ¯1 as the argument increases from 0 and that *A* is the smallest positive argument for which ¯1=*G A*. Show that *A* is a local minimum of *G*.

d. Show that as the argument *X* increases from *A*, if the values *G X* start to increase then they will continue to increase until, if ever, they reach 1.

e. Suppose that the values of *G* do reach 1 as the argument increases from *A* and that *B* is the smallest positive argument for which 1=*G B*. Show that *B* is a local maximum of *G*.

f. Sketch the graph of *G* on the interval (0≤ω)∧ω≤*B*.

g. We are now back to where we began. It is reasonable to expect that as the argument *X* increases from *B*, the graph of *G* looks exactly as it did when *X* increased from 0. Sketch the graph of *G* on the interval (*B*≤ω)∧ω≤2×*B*.

h. Sketch the graph of *G* on the interval (|ω)≤5×*B*. The function *G* is said to be <u>periodic</u> with <u>period</u> *B*.

i. Paraphrase the arguments of parts a-h for the case 0=*G* 0 and sketch the graph of the resulting function.

4.30 Using the function *T*4 in the text as a guide, for each function *F* and argument *A* below define the tangent function *TA* whose graph is the line tangent to the graph of *F* at the coordinate point *A*. Then evaluate *TA* at the indicated argument *X*.

a. *F*:ω⋆0.5; *A*←9; *X*←8.95
b. *F*:ω⋆÷3; *A*←27; *X*←27.02
c. *F*:1+ω⋆0.5; *A*←1; *X*←1.001

4.31 a. We say that two constants *A* and *B* agree to *N* decimal places or decimal digits if (0.5×10⋆-*N*)>|*A*-*B*. Which of the following pairs of constants agree to 3 decimal places?

i. 0.2571,0.25724
ii. 0.112 ,0.11199
iii. 0.8931,0.8924
iv. 0.8932,0.8965

Now round each constant to 3 decimal places and repeat the exercise.

b. Test the fact that if two constants *A* and *B* agree to *N* decimal digits then all the digits in *N*⍕|*A*-*B* are 0.

4.32 Use Newton's method to produce 6 decimal digit approximations to each of the constants defined below.

a. 3⋆0.5 b. 5⋆÷3 c. 7⋆2÷3

4.33 Newton's method has the property of <u>local convergence</u>, which means that a sequence produced by this method is guaranteed to approach a root *R* only if the sequence begins sufficiently close to *R*. For example, the polynomial with coefficient vector 0 10 ¯11 1 has the roots 0 1 10. If Newton's method for this polynomial is started at the argument 0.5 the resulting sequence will approach the root 10 rather then either of the roots 0 or 1. Trace this sequence geometrically by graphing the polynomial with coefficient vector 0 10 ¯11 1 and then graph the tangent line at the point 0.5,2.375. If *R*1,0 represents the point where this tangent line intersects the argument axis, graph the tangent line at the point *R*1, 0 10 ¯11 1 *POLY R*1. Continue this graphing procedure for the further approximations until the behavior of the sequence produced by Newton's method become apparent.

4.34 An accurate sketch of the graph of a function will suggest initial approximations to the roots of the function, and then more accurate approximations can be produced by Newton's method. For each vector *C* below sketch the graph of the polynomial *C POLY* ω on the interval 5≥|ω. Use the resulting approximations to the roots of the polynomial as initial approximations for Newton's method and produce accurate approximations to the roots. Then adjust your graph to take account of this more accurate information.

C←36 ¯21 ¯5 1
C←70 ¯71 ¯5 6
C←56 ¯19 ¯3 2
C←¯50 15 12 ¯4

4.35 A polynomial of degree 1 has exactly one root. Trace Newton's method geometrically (as in Ex 4.33) for a polynomial of degree 1. What do you see?

4.36 Graph the polynomial `0 5 0 ¯1 POLY ω` and geometrically trace the sequence produced by Newton's method, using `1` for the first approximation.

4.37 Repeat Ex 4.36 for the coefficient vector `1 0 1` and the initial approximation `2`.

4.38 Even though we have a good general purpose rootfinder in Newton's method we should not ignore the possibility of determining the roots of a function explicitly. Define a function which produces the roots of a quadratic function, that is, a polynomial of degree `2`.

4.39 Knowledge of critical arguments and their types is a useful aid in producing accurate sketches of graphs of monadic functions. Each vector *C* below is the coefficient vector of a polynomial which is to be graphed on the interval of arguments `5≥|ω`. First produce a sketch of the graph using the scheme of Ex 1.57. This sketch should indicate approximate locations of critical arguments. Use these approximate values and Newton's method to produce accurate approximations to the critical values and test their types. Finally, adjust your graph to take account of this more accurate information.

```
C←¯22.5 3.25 3
C←0.6 ¯79.2 ¯28.2 14
C←¯12 11 2
C←1 ¯27 12.6 ¯0.2 ¯1 0.16
```

4.40 Each function *F* below is accompanied by a subdivision *V* of the interval `(V[0]≤ω)∧ω≤V[¯1+ρV]` for which *F* is monotonic on each subinterval of *V*. Sketch the graph of *F* and evaluate `×DIFF F V`. Then use the strategy described in the text to sketch the graphs of the two associated monotonic functions.

```
a.  F:1+ω*2; V←¯2 0 3
b.  F:+/3 4 1×ω*1 2 3; V←0.5×ι5
c.  F:(20 ¯30 31 ¯10 1÷1 1 2 3 4) POLY ω; V←1+ι6
d.  F:0 0 ¯3 2 3 POLY ω; V←¯1 ¯0.5 0 0.5 1
```

4.41 a. For a vector *V* the value of the grade-up function `⍋V` is a permutation of order `ρV` for which the elements of `V[⍋V]` are in non-decreasing order. Describe the values `MP V` for argument vectors *V*, where `MP:(MPA ω)[⍋MPA ω]` and `MPA:ω,0.5×(1↓ω)+¯1↓ω`.

b. Repeat Ex 4.40 with the vectors *MP V* in place of *V* and describe the relation between the monotonic functions produced in this exercise and those produced in Ex 4.40.

c. Repeat part b with the vectors *MP MP V* in place of *MP V*.

4.42 If it is difficult to locate the local maxima and local minima of a function then it is also difficult to produce a subdivision *V* for which *F* is monotonic on each subinterval of *V*. Also, in rare instances *F* may oscillate infinitely often on an interval, in which case there is no such subdivision of that interval. Even so, it is usually possible to express *F* as the sum of two monotonic functions. Each function *F* below is accompanied by a subdivision *V* of the interval (*V*[0]≤ω)∧ω≤*V*[¯1+ρ*V*]; *F* is not necessarily monotonic on any of the subintervals of *V*. Consequently the associated functions *V RF W* and *V FF W* are not necessarily monotonic. However, the graphs of *V RF W* and *V FF W* can be produced by the strategy described in the text. Sketch the graph of *F*, evaluate ×*DIFF F V*, and then sketch the graphs of *V RF W* and *V FF W*. Do not be concerned with whether or not *F* is monotonic on each subinterval of *V*. Then repeat the exercise for *MP V* in place of *V*, then for *MP MP V* in place of *V*, and so on, where the function *MP* is defined in Ex 4.41. What do you observe?

a. *F*:×/ω-ι3; *V*←¯1+0.5×ι11
b. *F*:¯15 46 ¯36 3 *POLY* ω; *V*←0.2×ι16

4.43 <u>The second derivative test for the type of a critical argument.</u> Suppose that *A* is a critical argument of the function *F* and that <u>*DD*</u>*F* is the second derivative of *F*. What is the type of *A* if:

a. 1=×<u>*DD*</u>*F A* b. ¯1=×<u>*DD*</u>*F A* c. 0=×<u>*DD*</u>*F A*

4.44 Repeat Ex 4.4 using the second derivative test.

4.45 If *A*,*F A* is an inflection point of the function *F* then 0<<u>*DD*</u>*F X* for arguments *X* on one side of *A* and 0><u>*DD*</u>*F Y* for arguments *Y* on the other side of *A*. Therefore 0=<u>*DD*</u>*F A*. However, the fact that 0=<u>*DD*</u>*F B* does not necessarily mean that *B*,*F B* is an inflection point of *F*. For example, if *F*:ω*4 the coordinate point 0,0 is not an inflection point of *F* but 0=<u>*DD*</u>*F* 0. We can locate all inflection points of *F* by determining all arguments *A* for which 0=<u>*DD*</u>*F A* and eliminating those which the test in Ex 4.4 shows to be local maxima or local minima.

Determine all inflection points on the graph of the polynomial *C POLY ω* for each *C* below.

```
C←6 0 ¯36 1 0.5
C←24 ¯18 ¯3 1
C←8 4 ¯4 3 ¯1 0.1
C←15 ¯3 ¯10.5 ¯20.5 ¯0.6 0.1
```

4.46 The sketch of each graph in Ex 4.39 will suggest approximate locations of inflection points. Use these approximate locations and Newton's method to produce accurate approximations of the inflection points and correct the graphs accordingly.

4.47 For each vector *C* below determine all critical arguments and their types and all inflection points on the graph of the polynomial *C POLY ω*. Sketch a segment of the graph near each inflection point and each critical point. Use these segments and the points 5,*C POLY* 5 and ¯5,*C POLY* ¯5 to produce a sketch of the graph of the polynomial on the interval 5≥|ω.

```
C←3   12   2.5   ¯2   0.25
C←¯4   0   ¯45   ¯8   3.5   0.8
C←0   ¯64   24   ¯4   0.25
C←8   96   ¯2   0   1
```

4.48 The concept of concavity is useful in studying sequences produced by Newton's method. For example, draw the graph of a function which is everywhere increasing and concave upward and which has a root at *R*. Geometrically trace Newton's method (as in Ex 4.33) for this function using initial approximations to the left and right of *R* and show that after the first approximation, the approximations decrease towards *R*.

4.49 Draw the graph of a function which is everywhere decreasing, has a root *R*, and is concave downward to the left of *R* and concave upward to the right of *R*. Geometrically trace Newton's method for this function and show that the approximations alternate from one side to another about *R*.

4.50 Using the function *Q*4 in the text as a guide, for each function *F* and argument *A* of Ex 4.30 define the quadratic function *QA* which is a polynomial of degree 2 and such that (*F A*)=*QA A* and (*DF A*)=*DQA A* and (*DDF A*)=*DDQA A*. Then evaluate *QA* at the indicated argument *X*.

4.51 a. Show that 8.95*0.5 is a root of the function *G*:8.95-ω*2, and then use Newton's method to produce a six digit approximation of 8.95*0.5. Compare with the results of Exs 4.30a and 4.50a.

b. Show that 27.02*÷3 is a root of *G*:27.02-ω*3, and then use Newton's method to produce a six digit approximation to 27.02*÷3. Compare with the results of Exs 4.30b and 4.50b.

c. Define a polynomial for which 1+1.001*0.5 is a root and use Newton's method to produce a six digit approximation to 1+1.001*0.5. Compare with the results of Exs 4.30c and 4.50c.

4.52 Define the first through fourth order Taylor polynomials of *F* at *A* for each pair of a function *F* and an argument *A* in Ex 4.30. Evaluate the polynomials at the indicated argument *X* and compare these values to the answers to Ex 4.51.

4.53 Sketch a graph of the function ×/ω-1 2 3. Define the Taylor polynomials of this function at 1 of orders 1,2 and 3. Sketch graphs of these Taylor polynomials and compare the graphs to that of the function.

4.54 Repeat Ex 4.53 for the function +/0 ¯2 1 1×ω*0 1 2 3 and its Taylor polynomials at ¯1 of order 1,2 and 3.

4.55 Produce the coefficient vector *C* for which:

```
×/ω-1 2 3
+/C×(ω-1)*0 1 2 3
```

Compare the coefficient vectors defined by (*N*+1)↑*C* with the coefficient vectors of the Taylor polynomials of order *N* in Ex 4.53 when *N* is 1, 2 or 3.

4.56 Produce the coefficient vector *C* for which:

```
+/0 ¯2 1 1×ω*0 1 2 3
+/C×(ω+1)*0 1 2 3
```

Make the comparisons described in Ex 4.55 with Ex 4.54 in place of 4.53.

4.57 Suppose that the polynomial *C POLY* ω-*A* is the third order Taylor polynomial of the function *F* at the argument *A*. Show that the zeroth, first, second and third derivatives of *F* ω and *C POLY* ω-*A* are equal at the argument *A*.

4.58 Suppose that the polynomial *C POLY* ω-*A* is a Taylor polynomial of the function *F* at the argument *A*. Show for each integer *N* between 0 and ¯1+ρ*C* that the *N*th derivatives of *F* ω and *C POLY* ω-*A* are equal at the argument *A*.

4.59 Use the third order Taylor polynomial at *A* of the function ω*0.5 to produce an approximate value of each scalar defined below.

a. `9.1*0.5 ;A←9`
b. `4.01*0.5 ;A←4`
c. `8.9*0.5 ;A←9`
d. `.8*0.5 ;A←1`

4.60 Using the Times Rule for Taylor Coefficient Functions as a guide, define

```
TTIMES:(ρα)↑α TIMES ω
```

Then if `H ω` is equivalent to `(F ω)×G ω`:

```
α TH ω
(α TF ω) TTIMES α TG ω
```

Using expression 4.5.3 as a guide, define the Taylor coefficient functions of `F:ω*0.5`, `G:ω*0.25` and `H:ω×0.75`, and then test the above identity.

4.61 Use the Times Rule for Taylor Coefficient Functions to show that if `H ω` is equivalent to `B×F ω` for a scalar `B`, then `A TH ω` is equivalent to `B×α TF ω`.
Hint. Show that if `G:B+0×ω` then `TG:(α+1)↑B+0×ω`.

4.62 a. As in Ex 4.60, define the function `TPLUS` such that if `H ω` is equivalent to `(F ω)+G ω`, then `α TH ω` is equivalent to `(α TF ω)TPLUS α TG ω`.

b. Define the function `TCOMP` such that if `H ω` is equivalent to `F G ω`, then `α TH ω` is equivalent to `(α TF ω) TCOMP α TG ω`. Test your definition with the functions `F:ω*0.5`, `G:ω*0.25`, `H:ω*0.125`.

c. Define the function `TREC` such that if `H ω` is equivalent to `÷F ω`, then `α TH ω` is equivalent to `TREC α TF ω`. Test your definition with the function `F:1-ω`.

4.63 Define the Taylor coefficient function of each of the following functions.

a. `3×ω*÷3`
b. `(5+ω)*0.5`
c. `÷1-ω*2`
d. `(1 3 ¯1 POLY ω)÷1+ω*0.5`

4.64 Determine the value of the seventh derivative of `÷1+ω*0.5` for the argument `9`.

4.65 a. Sketch the graph of `4○ω`.
b. Define the Taylor coefficient function `TC4` of `4○ω`.

4.66 a. Sketch the graph of `¯4○ω`
b. Define the Taylor coefficient function `TCN4` of `¯4○ω`.

5

5.1 Which of the following vectors *T* populate the partition ¯1 1 4 5 7?

```
T←¯1 2 4 5              T←¯1 1 5 7
T←0 2.5 4.5 6           T←1 4 4.5 5
T←0 1 1.5 6
```

5.2 Repeat Ex 5.1 for the vectors *T* below and the partition 10 7 2 ¯1 ¯3.

```
T←7 2 ¯1 ¯3             T←9 5 1 0
T←10 2 0 ¯1             T←7 7 2 ¯2
```

5.3 Suppose that *P* is an increasing partition and that the vector *T* populates *P*. Show that ∧/*T*≥¯1↓*P* and ∧/*T*≤1↓*P*. If *P* is decreasing, show that ∧/*T*≥1↓*P* and ∧/*T*≤¯1↓*P*.

5.4 Sketch the region in a coordinate plane bordered by the graph of the function *F*:1+⍵⋆4, the horizontal axis, and the vertical lines at 0.5 and 2. Sketch the regions in the plane composed of rectangles whose areas are to the Riemann sums

```
a.   +/(F 1↓P)×DIFF P←0.5+0.5×⍳4
b.   +/(F ¯1↓P)×DIFF P←0.5+0.25×⍳7
c.   +/(F 0.5×(1↓P)+¯1↓P)×DIFF P←0.5+.1×⍳16
```

5.5 Repeat Ex 5.4 for the function *G*:16-⍵⋆2, the vertical lines at ¯4 and 4, and the Riemann sums

```
a.   +/(G 1↓P)×DIFF P←¯4+⍳9
b.   +/(G ¯1↓P)×DIFF P←¯4+0.5×⍳17
c.   +/(G 0.5×(1↓P)+¯1↓P)×DIFF P←¯4+⍳9
```

5.6 In the text we established one relation between difference-quotients and Riemann sums. Namely, for a partition *P* and a vector *T* that populates *P*:

```
(DIFF SUM (F T)×DIFF P)÷DIFF P
F T
```

Another relation between difference-quotients and Riemann sums is as follows. If *P*←*A*+*S*×⍳⍴*P* for scalars *A* and *S* then ∧/*S*=*DIFF P*. Consequently the expression (*DIFF F T*)÷*S* is a difference-quotient and the expression *SUM* (*F T*)×*S* is a vector of Riemann sums. Show that

```
(F T[0])+SUM ((DIFF F T)÷S)×S
F T
```

5.7 For distinct scalars *A* and *B* show that every vector of the form `A+((B-A)÷K)×⍳K+1` is a partition of the interval with end points *A* and *B*.

5.8 Consider the function `F:1+ω*2`. Each vector *P* below is an increasing partition of the interval `(0≤ω)∧ω≤2` and each corresponding vector *T* populates *P*. For each pair of vectors *P* and *T* use the test described in Ex 3.51 to indicate that the vector of Riemann sums `SUM (F T)×DIFF P` is a vector of values of a polynomial. What is the degree of the polynomial in each case?

```
P←0.5×⍳5   ; T← ¯1↓P
P←0.5×⍳5   ; T← 1↓P
P←0.2×⍳11  ; T←0.5×(1↓P)+¯1↓P
```

5.9 For each case of Ex 5.8, use the fitting polynomial procedure to fit the values `SUM (F T)×DIFF P` for arguments *P*.

5.10 Repeat Ex 5.8 for the function `F:ω+ω*3` and each of the following pair of vectors *P* and *T*.

```
P←1+0.5×⍳7    ; T←1↓P
P←1+0.25×⍳13  ; T←¯1↓P
P←1+0.5×⍳7    ; T←0.5×(1↓P)+¯1↓P
```

Note that each vector *P* is an increasing partition of the interval `(1<ω)∧ω≤4`.

5.11 Repeat Ex 5.9 for `F:ω+ω*3` and each case of Ex 5.10.

5.12 The vector *V* is said to be uniform if `0=DIFF DIFF V`. Suppose that *P* is a uniform partition of the interval `(A≤ω)∧ω≤B` and that *T* is a uniform vector that populates *P*. Use the identity

```
(DIFF SUM (C POLY T)×DIFF P)÷DIFF P
C POLY T
```

and the Fundamental Theorem of Finite Differences (Ex 3.50) to show that the vector of Riemann sums `SUM (C POLY T)×DIFF P` is the vector of values of a polynomial of degree `1+⍴C` for the vector of arguments *P*.

5.13 This exercise and Exs 5.14-15 establish the connection between the application discussed in Section 5.2 and Riemann sums. Show that the expression in the text for the approximate force on a side of a tank of water, namely,

+/DENSITY×L×Q×DIFF D, is a Riemann sum of the function *DENSITY×L×ω*.

5.14 Consider the function *F:M×ω* for a positive scalar *M*.

a. Draw the region bordered by the graph of this function, the horizontal axis, and the two vertical lines at *A* and *B*, where (0≤*A*)∧*A*<*B*.

b. Show that the area is 0.5×*M*×(*B*⋆2)-*A*⋆2.

c. Let *P* be an increasing partition of the interval (*A*≤ω)∧ω≤*B*. Draw the region composed of rectangles whose area is the value of the Riemann sum defined by the expression +/(*F* 0.5×(1↓*P*)+¯1↓*P*)×*DIFF* *P*. Devise a geometrical argument to show that

```
+/(F 0.5×(1↓P)+¯1↓P)×DIFF P
0.5×M×(B⋆2)-A⋆2
```

5.15 Use Exs 5.13 and 5.14 to show that

```
+/DENSITY×L×0.5×((1↓P)+¯1↓P)×DIFF P
0.5×DENSITY×L×-/D[(¯1+ρD),0]⋆2
```

Compare this identity with sequence 5.2.1.

5.16 The next group of exercises deals with another application of Riemann sums, the approximation of volumes of solids of revolution.

Sketch the half circle which is the graph of the function 0○ω. Shade the region bordered by the horizontal axis and the half circle. If this half disc is rotated about the horizontal axis it will sweep out a sphere.

Sketch the graph of *M*×ω, where *M* is a positive scalar. Shade the triangle bordered by this graph, the horizontal axis, and the vertical line at *B*, where *B*>0. If the triangle is rotated about the horizontal axis it will sweep out a cone.

Because of the way in which the sphere and the cone can be generated they are called <u>solids of revolution</u>. The volume of a solid of revolution can be approximated using Riemann sums as follows.

Suppose that *F* is a function for which 0≤*F* ω for all arguments in the interval (*A*≤ω)∧ω≤*B*. Let *P* be an increasing partition of this interval and let *T* be a vector that populates *P*. Sketch the region composed of rectangles whose area is equal to the value of the Riemann sum +/(*F T*)×*DIFF P*. If all the rectangles in the diagram are rotated about the argument axis, the resulting solid, which is composed of cylinders, has volume approximately equal to the volume of the solid of revolution obtained by rotating the graph of *F* about the argument axis. In order to describe the volumes of these cylinders we will use the <u>monadic circle</u> function denoted by ○ω, which is identical to *PI*×ω, where the value of *PI* is the circumference of a circle of diameter 1.

The volume of a cylinder with base radius *R* and altitude *H* is equal to ○*H*×*R*⋆2. Use this fact to show that the volume of the region composed of cylinders described above is equal to

```
○+/((F T)⋆2)×DIFF P
```

Define the function *G* for which the above expression is a Riemann sum. That is, define *G* for which

```
○+/((F T)⋆2)×DIFF P
+/(G T)×DIFF P
```

5.17 Suppose *R* is a positive scalar, *P* is an increasing partition of the interval *R*≥|ω, and *T* is a vector that populates *P*. Define a Riemann sum in terms of *P* and *T* whose value is approximately equal to the volume of a sphere of radius *R*.

5.18 Suppose *H* is a positive scalar, *P* is an increasing partition of the interval (0≤ω)∧ω≤*H*, and *T* is a vector that populates *P*. Define a Riemann sum in terms of *P* and *T* whose value is approximately equal to the volume of a circular cone of altitude *H* and base radius *R*.

5.19 Determine an increasing partition *P* of the interval (1≤ω)∧ω≤3 and a vector *T* that populates *P* for which

```
0.005≥|AREA-+/(F T)×DIFF P
```

where *F*:ω+2×ω⋆3 and the scalar called *AREA* is the area of the region bordered by the graph of *F*, the argument axis, and the vertical lines at 1 and 3.
Suggestion. Exercise 5.7 is helpful in determining a partition *P* of an interval (*A*≤ω)∧ω≤*B* for which the scalar defined by the expression ⌈/|*DIFF P* is less than or equal to a specified value.

5.20 Repeat Ex 5.19 for the interval (0≤ω)∧ω≤5 and *F*:10+ω.

5.21 Suppose that the scalar called *AREA* is the area of the region bordered by the graph of the function *SQ*:ω⋆2, the argument axis, and the vertical lines at 0 and 1. Determine a partition *P* such that for every vector *T* that populates *P*, 0.0005>|*AREA*-+/(*SQ T*)×*DIFF P*. According to this inequality, each of the Riemann sums +/(*SQ* 1↓*P*)×*DIFF P* and +/(*SQ* ¯1↓*P*)×*DIFF P* and +/((*SQ* 0.5×(1↓*P*)+¯1↓*P*)×*DIFF P* is an approximation of *AREA* accurate to two decimal places. However, one of these Riemann sums is a better approximation of *AREA* than the other two, even though this is not revealed by the above inequality. Which is it and why?

5.22 Inequality 5.3.2 is also valid for functions which are decreasing on an interval (*A*≤ω)∧ω≤*B*. Repeat Ex 5.19 for the interval (¯1≤ω)∧ω≤2 and the function *F*:8-ω⋆3.

5.23 Repeat Ex 5.19 for the interval 2≥|ω , the function *F*:0 0 0 7 0 ¯1 *POLY* ω and with .0005 in place of .01 in the left hand side of the inequality.

5.24 In this exercise we will estimate the value of scalar *AREA* which is the area of the region *R* bordered by the graph of the function *G*:9-ω⋆2, the argument axis, and the vertical lines at ¯1 and 1. Sketch this region. Divide this region into two parts, *R*1 being the part to the left of the vertical axis and *R*2 the part to the right. Let the scalars called *AREA*1 and *AREA*2 be the areas of the regions *R*1 and *R*2 respectively. Determine a partition *P*1 of the interval (¯1≤ω)∧ω≤0 and a vector *T*1 that populates *P*1 for which:

```
0.0025≥|AREA1-+/(G T1)×DIFF P1
```

Determine a partition *P*2 of the interval (0≤ω)∧ω≤1 and a vector *T*2 that populates *P*2 for which:

```
0.0025≥|AREA2-+/(G T2)×DIFF P2
```

Define *P*←*P*1,1↓*P*2 and *T*←*T*1,*T*2. Show that *P* is a partition of the interval (¯1≤ω)∧ω≤1, that *T* populates *P*, and that:

```
0.005≥|AREA-+/(G T)×DIFF P
```

5.25 Repeat Ex 5.19 for the interval (0≤ω)∧ω≤2 and the function 2 ¯2 1 *POLY* ω.

5.26 Repeat Ex 5.19 for the interval (0≤ω)∧ω≤3 and the function 0 12 ¯9 2 *POLY* ω.

5.27 In the preceding exercises you have determined Riemann sums whose values approximate the areas of certain regions within specified tolerances. In each case the required partition *P* has many elements and consequently the evaluation of the Riemann sum +/(*F T*)×*DIFF P* requires many evaluations of the function *F*. If *F* is a polynomial of small degree and if the vectors *P* and *T* are uniform, there is a procedure for evaluating the Riemann sum which requires only a few values of *F*. Consider the following example for the function *F*:ω⋆2:

```
P←.01×⍳1001
T←0.5×(1↓P)+¯1↓P
C←P[⍳4] CFP SUM (F T[⍳3])×DIFF P[⍳4]
```

```
      C POLY P[¯1+⍴P]               +/(F T)×DIFF P
333.33325                     333.33325
```

How many values of *F* were used to produce *C*? Why does this method of evaluating Riemann sums work?
Hint. See Ex 5.12.

5.28 If the vectors *P* and *T* determined in Ex 5.19 are not uniform, replace them with uniform vectors and use the method illustrated in Ex 5.27 to evaluate the resulting Riemann sum.

5.29 Repeat Ex 5.28 with Ex 5.20 in place of Ex 5.19.

5.30 Repeat Ex 5.28 with Ex 5.22 in place of Ex 5.19.

5.31 Repeat Ex 5.28 with Ex 5.23 in place of Ex 5.19.

5.32 Repeat Ex 5.28 with Ex 5.25 in place of Ex 5.19.

5.33 Repeat Ex 5.28 with Ex 5.26 in place of Ex 5.19.

5.34 Use Exs 5.17 and inequality 5.3.2 to determine the volume of a sphere of radius 2.5 to five decimal place accuracy.

5.35 Sketch the region composed of rectangles whose signed area is equal to the value of the following Riemann sum of `F:ω*3`:

```
      P←¯3 ¯2 ¯0.5 1 2.5
      T←¯3 ¯1 0 2
      +/(F T)×DIFF P
```

5.36 Repeat Ex 5.35 for the function `G:¯9+α*2` and the Riemann sum

```
      P←⍳10
      T←0.5×(1↓P)+¯1↓P
      +/(G T)×DIFF P
```

5.37 Inequality 5.3.2 can be derived for monotonic functions with possibly negative function values as follows. (As in the text of Section 5.3 we will present the proof for non-decreasing functions. You should construct the corresponding proof for non-increasing functions.) Suppose that the function *F* is non-decreasing on the interval `(A≤ω)∧ω≤B` and that the scalar called *SAREA* is the signed area of the region bordered by the graph of *F*, the argument axis, and the vertical lines at *A* and *B*. Define the function `G:(F ω)-F A`.

a. Show that *G* is a non-decreasing function on the interval (*A*≤ω)∧ω≤*B* and that 0≤*G* ω for all arguments ω in this interval. Let the scalar called *AREA* be the area of the region bordered by the graph of *G*, the argument axis, and the vertical lines at *A* and *B*. According to 5.3.2:

(|*AREA*-+/(*G T*)×*DIFF P*)≤|(-*G B*,*A*)×⌈/|*DIFF P*

b. Devise a geometric argument for the fact that *SAREA* equals *AREA*+(*F A*)×*B*-*A*.

c. Show that +/(*G T*)×*DIFF P* equals (+/(*F T*)×*DIFF P*)-(*F A*)×*B*-*A*.

d. Show that *AREA*-+/(*G T*)×*DIFF P* equals *SAREA*-+/(*F T*)×*DIFF P*.

e. Show that -/*G B*,*A* equals -/*F B*,*A*.

f. Finally, show that

(|*SAREA*-*A*/(*F T*)×*DIFF P*)≤|(-/*F B*,*A*)×⌈/|*DIFF P*

5.38 For *F*:25 0 ¯15 2 *POLY* ω determine an increasing partition *P* of the interval (1≤ω)∧ω≤4 and a vector *T* that populates *P* for which:

0.005≥|*SAREA*-+/(*F T*)×*DIFF P*

where *SAREA* is the signed area of the region bordered by the graph of *F*, the horizontal axis, and the vertical lines at 1 and 4.

5.39 Repeat Ex 5.38 for the interval (¯1≤ω)∧ω≤4 and the function *F*:(4×ω)-ω⋆2.

5.40 Repeat Ex 5.28 with Ex 5.38 in place of Ex 5.19.

5.41 Repeat Ex 5.28 with Ex 5.39 in place of Ex 5.19.

5.42 Sketch a graph of the function 9-ω⋆2 on the interval (0≤ω)∧ω≤5 and indicate the region bordered by the graph of this function, the horizontal axis, and the vertical lines at 0 and 5. Repeat this for the function defined by |9-ω⋆2. Note that the areas (not signed areas) of the two indicated regions are equal but only Riemann sums of |9-ω⋆2 approximate this area. Use 5.3.2 to estimate the area of these regions to within two decimal place accuracy.

5.43 Repeat Ex 5.42 for the function 3+(¯5×ω)+ω⋆2 and the interval (0≤ω)∧ω≤4.

5.44 Sketch the graphs of two functions *F* and *G* on the interval (*A*≤ω)∧ω≤*B* whose graphs intersect in at least one point. Shade the region bordered by the graphs of these functions and the vertical lines at *A* and *B*. Let *P* be an increasing partition of the interval (*A*≤ω)∧ω≤*B* and *T* a vector that populates *P*. Devise a geometric argument for the fact that the value of the following Riemann sum for the function *H*:|(*G* ω)-*F* ω

```
+/(H T)×DIFF P
```

is an approximation of the area of the shaded region in your diagram.

5.45 Use Ex 5.44 and Inequality 5.3.2 to estimate to within three decimal place accuracy the area of the region bordered by the graphs of the functions 4-ω⋆2 and ¯4+ω⋆2 and the vertical lines at ¯2 and 2.

5.46 Repeat Ex 5.45 for the functions α-α⋆2 and 2+α-α⋆2 and the vertical lines at ¯2 and 3.

5.47 It is worthwhile to experimentally explore the relationship between Riemann sums and integrals of polynomials. Suppose that *C* is a coefficient vector and *A* and *B* are scalars for which *A*<*B*. Suppose also that *P* is a uniform, increasing partition of the interval (*A*≤ω)∧ω≤*B* and that *T* is a uniform vector that populates *P*. Then the vector of Riemann sums *SUM* (*C POLY T*)×*DIFF P* is the vector of values of a polynomial of degree 1+*DEGREE C* for the vector of arguments *P*. If the vector named *RC* is the coefficient vector of the polynomial, then *RC* can be produced by a fitting polynomial procedure using the vector of arguments (2+*DEGREE C*)↑*P* and the vector of values (2+*DEGREE C*)↑*SUM* (*C POLY T*)×*DIFF P*, or equivalently *SUM* (1+*DEGREE C*)↑(*C POLY T*)×*DIFF P*. The value of the Riemann sum +/(*C POLY T*)×*DIFF P* is then equal to the function value *RC POLY B*.

Define *C*←1 5 ¯3, *A*←1 and *B*←4. For each value of ρ*P* specified below define the uniformly-spaced, increasing partition *P* of the interval (*A*≤ω)∧ω≤*B* such that ρ*P* is the specified value. Then define *T*←¯1↓*P*. Next, produce the coefficient vector *RC* described above and evaluate the Riemann sum +/(*C POLY T*)×*DIFF P* by evaluating *RC POLY B*. Compare the coefficient vector *RC* with the vector *D PINTEG C* for some choice of the scalar *D*. Can you guess the value of the signed area of the region bordered by the graph of the polynomial with coefficient vector *C*, the argument axis, and the vertical lines at *A* and *B*?

a. 100=ρ*P*
b. 500=ρ*P*
c. 1000=ρ*P*
d. 5000=ρ*P*
e. 1*E*5=ρ*P*

5.48 Repeat Ex 5.47 for each choice of *A*, *B* and *T* below.

a. A←1 ;B←4;T←1↓P
b. A←1 ;B←4;T←0.5×(1↓P)×¯1↓P
c. A←1 ;B←7;T←1↓P
d. A←¯3;B←2;T←¯1↓P

5.49 Repeat Exs 5.47 and 5.48 for each of the following coefficient vectors *C*.

a. C←1 2 1 ¯2 3   c. C←0 0 1
b. C←1 0 4 ¯7   d. C←4 0 ¯6 5 8 2

5.50 Use the Second Fundamental Theorem in the text to determine the value of the scalar called *AREA* in Ex 5.19.

5.51 Repeat Ex 5.50 for Ex 5.20 in place of Ex 5.19.

5.52 Repeat Ex 5.50 for Ex 5.22 in place of Ex 5.19.

5.53 Repeat Ex 5.50 for Ex 5.23 in place of Ex 5.19.

5.54 Repeat Ex 5.50 for Ex 5.24 in place of Ex 5.19.

5.55 Repeat Ex 5.50 for Ex 5.25 in place of Ex 5.19.

5.56 Repeat Ex 5.50 for Ex 5.26 in place of Ex 5.19.

5.57 Use the Second Fundamental Theorem to determine the volume of the sphere in Ex 5.34.

5.58 Use the Second Fundamental Theorem and Ex 5.17 to show that the volume of a sphere of radius *R* is (○4÷3)×*R*⋆3.

5.59. Use the Second Fundamental Theorem and Ex 5.18 to show that the volume of a circular cone of altitude *H* and base radius *R* is (○÷3)×*H*×*R*⋆2.

5.60 Use the Second Fundamental Theorem to determine the value of the scalar called *SAREA* in Ex 5.38.

5.61 Repeat Ex 5.60 with Ex 5.39 in place of Ex 5.38.

5.62 Use the Second Fundamental Theorem to determine the area of the regions described in Ex 5.42.

5.63 Repeat Ex 5.62 with Ex 5.43 in place of Ex 5.42.

5.64 Repeat Ex 5.62 with Ex 5.45 in place of Ex 5.42.

5.65 Repeat Ex 5.62 with Ex 5.46 in place of Ex 5.42.

5.66 In the text we proved 5.4.6 when the integer *N* is 1. In the next three exercises we will prove another special case of 5.4.6, when the integer *N* is 2.

We will first verify the analog of the sequence 5.4.7 in the text. Verify the following sequence of identities for the Riemann sum defined by 5.4.5 when *N* is 2:

```
S×+/P S×⍳K
S×+/(S×⍳K)*2
(S*3)×+/(⍳K)*2
(D*3)×(+/⍳K)*2)÷K*3
```

5.67 Next, we prove the analog of the sequence 5.4.8 in the text. Use the identity

```
+/(⍳K)*2
(÷6)×K×(¯1+K)×¯1+2×K
```

to prove

```
+/(⍳K)*2
+/(0 1 ¯3 2÷6)×K*0 1 2 3
```

5.68 Use Exs 5.66 and 5.67 to prove:

```
S×+/P S×⍳K
(D*3)×+/(0 1 ¯3 2÷6)×K*¯3 ¯2 ¯1 0
```

Then argue for the fact that as the integer *K* is assigned large values, the latter expression in the above identity assumes values close to `(D*3)÷3`. Conclude that Identity 5.4.6 is valid when the integer *N* is 2.

We will now prove identity 5.4.6 for an arbitrary non-negative integer *N*.

5.69 We will first prove the analog of the sequence 5.4.7 in the text for an arbitrary non-negative integer *N*. Prove for the function `P:ω*N` and the Riemann sum defined by expression 5.4.5 that

```
S×+/P S×⍳K
(D*N+1)×(+/(⍳K)*N)÷K*N+1
```

5.70 To prove the analog of the sequence 5.4.8 in the text we must show that there is a coefficient vector of *CN* of degree *N*+1 for which:

```
+/(⍳K)*N
+/CN×K*⍳N+2
```

Assuming that there is such a coefficient vector, use the above identity and Ex 5.69 to prove:

```
S×+/F S×⍳K
(D*N+1)×+/CN×K*-⌽⍳K+2
```

Now argue for the fact that as *K* is assigned large values, the values of the second expression in this identity approach the value `(D*N+1)×CN[¯1+⍴CN]`. Finally, argue for the fact that identity 5.4.6 in the text is valid if `CN[¯1+⍴CN]=÷N+1`.

5.71 We can prove the existence of the coefficient vector *CN* described in Ex 5.70 by examining vectors of values of the function of *K* defined by `+/(⍳K)*N`. Show that

```
DIFF SUM (⍳K)*N
((N⍴0),1) POLY ⍳K
```

and use the Fundamental Theorem of Finite Differences (Ex 3.50) to conclude that there is a coefficient vector *CN* of degree *N*+1 for which:

```
+/(⍳K)*N
+/CN×K*⍳N+2
```

Use the fitting polynomial procedure to test the equation `CN[¯1+⍴CN]=÷N+1`.

5.72 For a vector *V* the scalar defined by the expression `(+/V)÷⍴V` is called the mean or average of the elements of *V*. Let `P←A+(⍳N+1)×(B-A)÷N` for scalars *A* and *B* and suppose that *T* is a vector that populates *P*. For a function *F*, the scalar defined by `(+/F T)÷⍴F T` is the mean of the vector of values *F T*. Verify the following sequence of identities:

```
(+/F T)÷⍴F T
(+/F T)÷⍴T
(+/F T)÷N
(÷B-A)×(+/F T)×(B-A)÷N
(÷B-A)×(+/F T)×DIFF P
```

This sequence of identities shows the relation between the means of vectors of values of a function *F* and Riemann sums of *F*. Moreover, this sequence of identities suggests that we define the mean of the values of *F* on the interval with end points *A* and *B* to be the scalar defined by the expression

```
(÷B-A)×A S̲F B
```

where *S̲F* is the definite integral function of *F*.

5.73 For each vector *C* and pair of scalars *A* and *B* below determine the mean of the values of the polynomial with coefficient vector *C* on the interval with end points *A* and *B*.

```
a.  C←1 3 ¯2 1;A←0 ;B←4
b.  C←0 0 0 1 ;A←¯2;B←2
c.  C←2 ¯4 3   ;A←4 ;B←¯1
```

5.74 For a vector *C* and a scalar *Y* a solution *X* of the equation *Y=C POLY* ⍵ can be produced by a rootfinder. Precisely, *X* is a zero of (*C*-(⍴*C*)↑*Y*) *POLY* ⍵. To see that this is true verify the following sequence of identities:

```
Y=C POLY X
0=(C POLY X)-Y
0=(C POLY X)-((⍴C)↑Y) POLY X
0=(C-(⍴C)↑Y) POLY X
```

5.75 For a vector *C* and a pair of distinct scalars *A* and *B* the expression (÷*B*-*A*)×(*A*,*B*) *DEFINT C* yields the mean of the values of the polynomial with coefficient vector *C* on the interval with end points *A* and *B*. It is true that there is at least one argument *M* in this interval for which:

```
C POLY M
(÷B-A)×(A,B) DEFINT C
```

The function value *C POLY M* is called the mean value of the polynomial with coefficient vector *C* on the interval with end points *A* and *B*. Use Newton's method to determine such an argument *M* for each vector *C* and pair of scalars *A* and *B* in Ex 5.73.

5.76 Use the trapezoid method to approximate the value of the scalar defined by *P*[0,¯1+⍴*P*] *DEFINT C* for each vector *C* and partition *P* below. In each case compare your answer to *P*[0,¯1+⍴*P*] *DEFINT C*.

```
a.  C←5 ¯3 2 1; P←4+0.1×⍳11
b.  C←¯2 4 2; P←3-0.2×⍳9
c.  C←0 ¯3 7 ¯1 0 2; P←¯2+0.5×⍳15
```

5.77 Repeat Ex 5.76 for Simpson's method in place of the trapezoid method.

5.78 The approximation method that uses the vector *COEFF* ⍳2, that is 0.5 0.5, is called the trapezoid method for the following reason. Sketch the graph of a function *F* whose values are positive. Choose scalars *X* and *S* for which *S*>0 and plot the coordinate points *X*,0 and *X*,*F X* and (*X*+*S*),0 and (*X*+*S*),*F X*+*S*. Sketch the trapezoid with these four coordinate points as vertices and show that the area of the trapezoid is identical to *S*×0.5 0.5+.×*F X*,*X*+*S*.

5.79 For a monadic scalar function *F* and a partition *P* for which there are scalars *A* and *S* such that *P*=*A*+*S*×⍳⍴*P*, the trapezoid method can be defined by the expression *S*×+/0.5 0.5+.×*F P*[(⍳2)∘.+⍳¯1+⍴*P*]. Show that

```
S×+/0.5 0.5+.×F P[(⍳2)∘.+⍳¯1+⍴P]
+/(0.5×(F 1↓P)+F¯1↓P)×DIFF P
```

5.80 a. In determining volumes of revolution of a function *F* we began with an approximating sum which could be written as a Riemann sum of another function *G*. Thus the volumes of revolution of *F* are values of the definite integral of *G*. We will now examine <u>lengths</u> of graphs, which provide another example of the importance of writing approximating sums as Riemann sums.

To analyze lengths of graphs we will use the <u>Mean Value Theorem</u>, which states that if an interval (*A*≤ω)∧ω≤*B* is in the domains of both a function *F* and its derivative *D̲F*, then there is an argument *T* in this interval for which:

```
((F B)-F A)=(D̲F T)×B-A
```

It is easy to devise a geometric argument for the validity of this theorem. For example, sketch the graph of the function *F*:9-ω⋆2 on the interval (¯2≤ω)∧ω≤1, and also sketch the secant line joining the coordinate points ¯2,*F* ¯2 and 1,*F* 1. Next, place a ruler so that it is tangent to the graph at ¯2,*F* ¯2. As the argument slowly increases from ¯2 to 1 move the ruler so that it is continually tangent to the curve. Stop when the ruler is parallel to the drawn secant line, and let *T* represent the argument for which *T*,*F* *T* is the coordinate point at which you are stopped. Evidently the slope of the ruler is *D̲* *T*, and this is equal to the slope of the secant line, which is -/(*F* 1 ¯2)÷-/1 ¯2. Consequently (*F* 1)-*F* ¯2 is equal to (*D̲F* *T*)×1-¯2.

b. To analyze lengths of graphs, sketch a graph of a function *F* over an interval (*A*≤ω)∧ω≤*B*. Choose an increasing partition *P* of this interval and draw a polygonal curve by connecting each pair of coordinate points *P*[*I*],*F* *P*[*I*] and *P*[*I*+1],*F* *P*[*I*+1] with a straight line. Show that the length of this polygon is

```
+/(((DIFF P)⋆2)+(DIFF F P)⋆2)⋆0.5
```

Next, use the Mean Value Theorem to show that there is a vector *TP* that populates *P* for which

```
∧/(DIFF F P)=(D̲F TP)×DIFF P
```

Then verify the following sequence of identities:

```
+/(((DIFF P)⋆2)+(DIFF F P)⋆2)⋆0.5
+/(((DIFF P)⋆2)+((D̲F TP)⋆2)×(DIFF P)⋆2)⋆0.5
+/((1+(D̲F TP)⋆2)⋆0.5)×DIFF P
+/(4○D̲F TP)×DIFF P
```

Evidently this expression is a Riemann sum of the function *G*:4○*D̲F* ω, and consequently its value is approximately *A* *S̲G* *B*. Also, the lengths of the above polygonal curves approach the lengths of the graph of *F* between the coordinate points *A*,*F A* and *B*,*F B* as the partition *P* is chosen so that ⌈/|*DIFF P* is small. Thus the length of this graph is *A* *S̲G B*.

c. For each function *F* below define the approximate definite integral function *S̲G* of *G*:4∘*D̲F ω* as in the text. Use the function *S̲G* to produce an accurate approximation of the length of the graph of *F* between the coordinate points *A*,*F A* and *B*,*F B*. For example, if *F*:(1-*ω*⋆2)⋆0.5 and *B*←0.5×2⋆0.5 then the length of the graph of *F* between the coordinate points 0,*F* 0 and *B*,*F B* is ○÷4. This value can be approximated as follows (the derivative of *F* is *D̲F*:-*ω*÷(1-*ω*⋆2)⋆0.5):

```
      0 S̲G B
0.7853981659
      ○÷4
0.7853981634
```

i. *F*:9-*ω*⋆2 ;*A*←¯3 ;*B*←3
ii. *F*:1+3×*α* ;*A*←0 ;*B*←5
iii. *F*:1 3 0 1 *POLY ω*; *A*←¯1; *B*←2

5.81 The Mean Value Theorem of Ex 5.80 can be used to prove the fact concerning integrals stated in Section 1.7, namely, that if the domain of a function *F* is an interval, say, (*A*≤*ω*)∧*ω*≤*C*, and if *G* and *H* are integrals of *F*, then there is a constant *S* for which *G ω* is identical to *S*+*H ω*. To prove this, consider the function *K*:(*G ω*)-*H ω*. Show that *D̲K*:0. Next, according to the Mean Value Theorem, for each scalar *B* in the interval there is another scalar *T*, between *A* and *B*, for which (*K B*)-*K A* is equal to (*D̲K T*)×*B*-*A*. Show that this result implies that *K B* is equal to *K A*, and therefore that there is a constant *S* for which *S*=*K B* for all *B* in the interval. Conclude that *G ω* is equivalent to *S*+*H ω*.

5.82 The procedure for approximating definite integrals will not be effective for every function, but only for those that can be closely approximated by fitting polynomials of fairly small degree. For example, sketch the graph of the function *ω*⋆0.5 on the interval (0≤*ω*)∧*ω*≤1, showing that the graph has a vertical tangent line at the coordinate point 0 0. No polynomial has a vertical tangent line, and therefore *ω*⋆0.5 cannot be closely approximated near 0 by fitting polynomials of fairly small degree. Define the function *S̲BF* of *F*:*ω*⋆0.5 as in the text. Compare the values 0 1 *S̲BF K* for various values of *K*.

6

6.1 The proofs of identities 6.1.1 can be modeled after those of the analogous identities for *SUM* and *DIFF* of

Section 1.1. However, instead of proving these identities for another particular pair of functions we can examine the general relation between scan and "pairwise". For primary dyadic scalar functions f and g define the functions called *F* and *G* according to the identities

```
F X                            C G X
(1↓X)f¯1↓X                     g\C,X
```

for scalars *C* and vectors *X*. Trace the proofs in the Exs 1.16-19 with f and g in place of - and + respectively to determine the necessary characteristics of f and g (e.g., commutativity, associativity, etc.) so that:

```
X                              Y
F C G X                        Y[0] G F Y
```

6.2 Show that `QT A*⍳N+1` equals `N⍴A` for the following three cases:

```
A←4; N←6                       A←1.5; N←8
A←¯3; N←5
```

6.3 a. Show that `QT C×X*⍳⍴C` equals `X×QT C` for the following three cases:

```
C←1 2 4 6; X←3                 C←!⍳10; X←2
C←3 ¯6 2 5; X←¯5
```

b. Show that `S×PRD N⍴A` equals `S×A*⍳N+1` for the following three cases:

```
S←5; A←3; N←7                  S←¯2; A←÷3; N←4
S←1.5; A←¯2.5; N←6
```

6.4 As a second application of Theorem 6.1.2 consider the problem of producing four decimal place approximations of the values of `(1+⍳100) POLY X` for all arguments *X* such that `0.5≥|X`. As in the example in the text, the problem reduces to producing an integer *L* for which

```
0.00005>|(¯100↑L↓1+⍳100) POLY X
```

whenever `0.5≥|X`. As before, we substitute `1+⍳100` for *C* in the statement of the theorem; we will substitute `0.5` for *B* later. Evaluate `QT 1+⍳N` for `N←3; N←5; N←8`. Derive an equivalent expression for `B×⌈/QT K↓|1+⍳100` analogous to the last expression in 6.1.3. The value of the scalar *M* in the statement of the theorem is identical to value of the derived expression. Show that `1>M` if `B<(K+1)÷K+2`. Next, produce an equivalent expression for `(|C[K]×B*K)÷1-M` analogous to the last expression in 6.1.4.

6.5 Substitute 0.5 for *B* in the last expression produced in Ex 6.4 and then determine the integer *L* described in that exercise.

6.6 Show that the above approximation procedure applies to any of the polynomials with coefficient vectors 1+ι*N*.

6.7 Show that the above approximation procedure can be used to produce approximate values of any specified accuracy of the polynomials with coefficient vectors 1+ι*N* for the interval of arguments 0.5≥|ω.

6.8 Unlike the example in the text, the above approximation procedure does not apply to all intervals of the form *B*≥|ω. Show that the procedure applies if and only if *B*<1.

6.9 We will now examine another approach to the polynomials with coefficient vectors 1+ι*N*. Use the function *REC* of Section 3.5 to show that the *N*th order quotient of 1 ¯2 1 is 1+ι*N* and the remainder is (-*N*+3)↑(*N*+2),-*N*+1. That is, test the identity

```
÷(1-ω)*2
((1+ιN) POLY ω)+(((N×2)×ω*N+1)-(N+1)×ω*N+2)÷(1-ω)*2
```

6.10 The values of the expression *X*⋆*N* approach 0 as the integer *N* is assigned large values if and only if 1>|*X*. Use this fact and Ex 6.9 to show that the values of the polynomials with coefficient vectors 1+ι*N* approach the values of the expression ÷(1-*X*)⋆2 as *N* is assigned large values if and only if 1>|*X*.

6.11 Use Exs 6.6 and 6.10 to show that for all arguments *X* such that 0.5≥|*X* and for the integer *L* determined in Ex 6.5:

```
0.00005>|((1+ιL) POLY X)-÷(1-X)*2
```

6.12 As a third example of Theorem 6.1.2, consider the problem of producing five decimal place approximations of (¯0.5⋆ι50) *POLY X* for all arguments *X* such that 1≥|*X*. Repeat Ex 6.4 with |¯0.5⋆ι*N* in place of 1+ι*N*.

6.13 Repeat Ex 6.5 with Ex 6.12 in place of Ex 6.4 and with 1 substituted for *B*.

6.14 Repeat Ex 6.6 for the above approximation problem.

6.15 Repeat Ex 6.7 for the above approximation problem.

6.16 Repeat Ex 6.8 for the above approximation problem and with *B*<2 in place of *B*<1.

6.17 Use the polynomial reciprocal function *REC* to test the identity

```
÷1+0.5×ω
((¯0.5*ιN) POLY ω)+((0.5×ω)*N)÷1+0.5×ω
```

6.18 Use the identity in Ex 6.17 to show that (¯0.5*ι*N*) *POLY X* approaches ÷1+0.5×*X* as *N* is assigned large values if and only if 2>|*X*.

6.19 Show that for all arguments *X* such that 1≥|*X* and for the integer *L* of Ex 6.13:

```
0.5E¯5>|((¯0.5*ιL) POLY X)-÷1+0.5×X
```

The point of the preceding exercises is that the polynomials with coefficient vectors 1+ι*N* and ¯0.5*ι*N* approximate certain known functions on specified intervals of arguments, and it is precisely on these intervals that the Theorem 6.1.2 can be applied. One might ask why this theorem is used to estimate the errors in these approximations when the remainders produced from the function *REC* can be used for that purpose. The reason for concentrating on the theorem is that it is not necessary to know a rule for the function being approximated in order to estimate how close the values of the approximating polynomials are to the values of the function. In this way when polynomials are used to construct new functions, values of the new function can be produced to within any specified tolerance.

6.20 Use 4.5.3 to define the Taylor coefficient function *T̲S* of *S*:ω*0.5. Then use Theorem 6.1.2 to show that the Taylor polynomials (*N T̲S A*) *POLY* ω-*A* provide approximate values of *S* ω on the interval (|*A*)>|ω-*A*.

6.21 Determine an integer *N* for which the Taylor polynomials (*N T̲S* 4) *POLY* ω-4 provide 10 digit approximations to the values of *S* ω on the interval 2≥|ω-4, where *S* and *T̲S* are the functions in Ex 6.20.

6.22 In the text we showed for the vector *C* for which identity 6.2.2 is valid, that (*R*×*C*[*I*])=*C*[*I*+1]×*I*+1. Show that these equations are equivalent to the identity *QT C* equals *R*÷1↓ιρ*C*. If *C*[0]=*C*0, use Identities 6.1.1 to show that *C*0×*PRD R*÷1↓ιρ*C* equals *C*. Finally, use the definition of the function *PRD* to show that *C*0××\1,*R*÷1↓ιρ*C* equals *C*. Compare the second expression of this identity with 6.2.3.

6.23 Prove Identity 6.2.4.

6.24 Show that

```
((E A)÷!⍳N) POLY ⍵-A
(E A)×(÷!⍳N) POLY ⍵-A
```

According to this identity and the definition of the Taylor coefficient function *TE*, the values (*E A*)×(÷!⍳*N*)*POLY X-A* approach the value *E X* as *N* is assigned large values. Evidently, according to the formal definition of the function *E*, those values also approach (*E A*)×*E X-A*. Consequently *E X* equals (*E A*)×*E X-A*. This addition formula for *E* can be phrased in the form we have used previously by substituting ⍵+⍺ for *X* and ⍵ for *A*:

```
E ⍵+⍺
(E ⍵)×E ⍺
```

6.25 Test the addition formula for the function *E* using the approximate values given in Table 6.1.7. For example, the approximate values for the arguments 0.2, 0.3 and 0.5 are 1.221, 1.350, and 1.649 respectively. Compare 1.649 and 1.221×1.350.

6.26 Use Table 6.1.7 to sketch the graph of the function *E* on the interval (0≤⍵)∧⍵≤2.

6.27 The addition formula for *E* can be used to extend Table 6.1.7. For example, if each element in the table is multiplied by the last element in the table, the result is a table for *E* of (approximately) two decimal place accuracy on the interval (2≤⍵)∧⍵≤4. Why? Describe a procedure for extending that table to a table of approximate values of *E* on the interval 5≥|⍵. The resulting table is accurate to approximately one decimal place. According to Table 6.1.9, what is the smallest value of *K* for which (÷!⍳*K*) *POLY* ⍵ will produce one decimal place approximations to values of *E* on the interval 5≥|⍵?

6.28 The procedure described in Ex 6.27 can also be used to extend the sketch of the graph of *E* produced in Ex 6.26. For example, a sketch of the graph on the interval (2≤⍵)∧⍵≤4 can be produced simply by scaling the graph in Ex 6.26 by the scale factor *E* 2 or approximately 7.389. Use this procedure to sketch the graph on the interval 4≥|⍵. Compare this sketch with the one produced in Ex 4.27.

6.29 Show that for each argument *X*, the line passing through the coordinate points *X*,*E X* and (*X*-1),0 is the tangent line to the graph of *E* at the coordinate point *X*,*E X*.

6.30 Show that if *R* and *RNEXT* are successive approximations to a root of the function *E* produced by Newton's method, then *RNEXT*=*R*-1. Find all the zeros of *E*.

6.31 On which intervals is the function *E* positive in value? Negative? Increasing in value? Decreasing? Concave upward? Concave downward?

6.32 Use the sketch of the graph of the function *E* produced in Ex 6.28 to produce the graph of each of the following functions.

a.	*E* 6×ω	d.	*E*-ω
b.	*E* 0.5×ω	e.	4×*E* ω÷¯3
c.	3×*E* 2×ω	f.	¯2×*E* ¯0.5×ω

6.33 If a table of approximate values of *E* is available then approximate values at intermediate arguments can be produced using a Taylor polynomial of *E*. For example, consider Table 6.1.7, whose entries are three decimal digit approximations of values of *E*. For every argument *T* such that (0≤*T*)∧*T*≤2, *T* is equal to *X*+*S*, where *X* is an argument in that table and 0.5≥|*S*. Since 0.0005>0.05 *EST* 3, the second order Taylor polynomial 1+*S*+0.5×*S*⋆2 will produce a 3 decimal digit approximation to the value *E S* if 0.05≥|*S*. According to the addition formula for the function *E*, the value

(*E X*)×1+*S*+0.5×*S*⋆2

is then an approximation to the function value *E T*. As an example, to produce an approximation to *E* 1.52, write down 3 copies of the approximate value of *E* 1.5 from the table in one column and the terms 1, .02, 0.5×.02⋆2 in another column. Form the products of the columns to produce a new column and sum the elements in the new column to produce the approximate value of *E* 1.52. Use Table 6.1.7 and this procedure to produce approximations to each of the following values.

a.	*E* 0.82	d.	*E* 0.612
b.	*E* 1.33	e.	*E* 0.78
c.	*E* 1.69	f.	*E* 1.45

6.34 The monadic function ⌊ (called floor) produces for a scalar *X* the largest integer whose value does not exceed *X*. For example, 3=⌊3.14 and ¯4=⌊¯3.14. Test the inequalities

(0≤*X*-⌊*X*)∧1>*X*-⌊*X*

for scalars *X*.

6.35 Use the addition formula for *E* to show that

E X
(*E* ⌊*X*)×*E X*-⌊*X*

6.36 Use an induction argument to show for positive integers *N* that (*E* 1)⋆*N* is equivalent to *E N*. First show that the identity is valid when *N* is 1. Then assume that it is valid for some specified value of *N*, to be called *K*, and show that it is then valid for *K*+1 by verifying the following sequence of identities:

```
E K+1
(E 1)×E K
(E 1)×(E 1)*K
(E 1)*K+1
```

6.37 Use the addition formula for *E* to show that *E -X* equals ÷*E X*, and equivalently *E -X* equals (*E X*)*¯1. In particular *E* ¯1 equals (*E* 1)*¯1.

6.38 Use an induction argument to show for negative integers *N*:

```
E N
(E 1)*N
```

6.39 Use Exs 6.35-6.38 to show that ((*E* 1)*⌊*X*)×*E X*-⌊*X* equals *E X*.

According to this identity all values of *E* can be produced from values for arguments in the interval (0≤⍵)∧⍵≤1. Also, according to the function called *EST* defined in Section 6.1, the polynomial with coefficient vector ÷!⍳14 will produce 10 decimal digit approximations to the values of *E* on the interval (0≤⍵)∧⍵≤1. These facts suggest the following practical definition of the function *E*:

```
EP:((+/÷!⍳14)*⌊⍵)×(÷!⍳14) POLY ⍵-⌊⍵
```

For arguments in the interval (0≤⍵)∧⍵≤1, the values of *EP* are 10 decimal digit approximations to the values of *E*. In general, the values of *EP* are approximately 10 digit approximations to the values of *E*.

6.40 In the text we showed that if *Y* is a vector for which Identity 6.2.6 is valid then *Y*[*I*+1] equals *Y*[*I*]×1+*R*×(*DIFF X*)[*I*]. Show that this equation is equivalent to the identity:

```
QT Y
1+R×DIFF X
```

If *Y*[0]=*Y*0, use 6.1.2 to show that *Y*0×*PRD* 1×*R*×*DIFF X* equals *Y*, or equivalently that *Y*0××\1,1+*R*×*DIFF X* equals *Y*.

6.41 We have already encountered vectors of values of growth and decay functions in Chapter 1. Repeat Exs 1.31 through 1.34, using expression 6.2.8 to produce the required vectors *Y*.

6.42 Suppose that interest on the principal on deposit in a bank is compounded at the rate *R* every *T* years. Then if the scalar *P* represents the principal at the beginning of an interest period, the principal at the end of the interest period is *P+P×R×T*. Show that if *P0* represents the initial deposit and the new principal is recorded at the end of *N* interest periods with no new deposits being made, the entries in the bankbook can be produced by the expression

```
P0××\1,1+R×DIFF T×⍳N+1
```

6.43 Use Ex 6.42 to produce the entries in a bankbook for an initial deposit of 1000 dollars at the interest rate 0.05 compounded annually for 10 years.

6.44 Repeat Ex 6.43 with compounded semi-annually (that is, every six months) in place of compounded annually.

6.45 Repeat Ex 6.43 with interest compounded quarterly (that is with *T*=0.25) instead of annually.

6.46 Suppose that *PR* is the vector defined in Ex 6.42, that is *PR*←*P0*××\1,1+*R*×*DIFF* *T*×⍳*N*+1. Arguing as in the text, as the interest period *T* is diminished to zero, the vector *PR* is approximately equal to the vector of values *P0*×*E* *R*×*T*×⍳*N*+1. Interest is said to be <u>compounded continuously</u> at the rate *R* if the initial deposit *P0* grows to *P0*×*E* *R*×*X* dollars after *X* years. Using Table 6.1.7 to produce approximate values of *E*, repeat Ex 6.43 with compounded continuously in place of compounded annually.

6.47 Repeat Exs 6.43-46 with an interest rate of 0.06.

6.48 According to Exs 6.43-47, which has the greater effect on the growth of principal, the interest rate or the interest period?

6.49 Bacteria multiply by division, so that in a well-fed colony the bacteria increase by a factor *R* of the number present. Precisely, if the function called *N* is defined so that the function value *N T* is equal to the number of bacteria present at time *T*, then *N T+S* is approximately equal to (*N T*)+(*N T*)×*R*×*S*. Also, *N* can have only integer values. Define *N*:*N0*×⌊*E* *R*×⍵-*T0*, where *N0* represents the population at time *T0*. Test the facts that *N0*=*N T0* and if *N0* is large, then *N T+S* is approximately (*N T*)×1+*R*×*S*.

6.50 Suppose that the number of bacteria in a particular colony increase by thirty per cent per unit of time, and that at time 2 their number is approximately 1*E*9. Use Ex 6.49 and Table 6.1.7 to estimate their number at time *T* for each *T* below.

a. *T*←4 b. *T*←7 c. *T*←0

6.51 Show for scalars *A*,*R* and *T*0 and the function *F*:*A*×⋆*R*×⍵-*T*0 that

```
D̲F T
R×F T
```

and *A*=*F T*0.

6.52 Use the primary monadic scalar function ⋆⍵ to define the function whose value at time *T* is identical to the number of dollars on deposit in a bank at time *T*, provided that interest is compounded continuously at the interest rate *R* and that *P*0 dollars are on deposit at time *T*0.
Hint. See Ex 6.46.

6.53 How much money must be invested at the interest rate of 5 per cent compounded continuously if the principal is to grow to 10000 dollars in 8 years?

6.54 A capacitor of 2*E*¯5 farads is connected to a source of electromotive force of 1000 volts and a resistor of 2000 ohms, as illustrated in Section 6.2. Assuming that *T*←0 is the instant when the switch is closed, define the function of *T* which describes the charge on the capacitor at time *T*. Sketch the graph of this function and describe its behavior for large values of *T*.

6.55 Repeat Ex 6.54 for a capacitor of 1*E*¯3 farads, a source of 10000 volts and a resistor of 4000 ohms, assuming that *T*←10 is the instant when the switch is closed.
Hint. See Ex 6.51.

6.56 There are many identities involving the circular functions, and many of these can be verified in terms of the geometric interpretation of pairs of values of *COS* and *SIN* as coordinates of points on the unit circle. For example, ((*COS* ⍵)⋆2)+(*SIN* ⍵)⋆2 equals 1. As another example, sketch several pairs of coordinate points (*COS X*), *SIN X* and (*COS* (○1)-*X*), *SIN* (○1)-*X*. Evidently:

```
COS (○1)-⍵               SIN (○1)-⍵
-COS ⍵                   SIN ⍵
```

Use geometric arguments to verify the following identities.

```
a.  COS (○1)+⍵           SIN (○1)+⍵
    -COS ⍵               -SIN ⍵

b.  COS ⍵                SIN ⍵
    COS-⍵                -SIN-⍵

c.  COS ⍵                SIN ⍵
    SIN (○÷2)-⍵          COS (○÷2)-⍵
```

d. *COS* ω   *SIN* ω
    SIN (o÷2)+ω   -*COS*(o÷2)+ω

e. *COS* (o÷4)-ω   *SIN* (o÷4)-ω
    SIN (o÷4)+ω   *COS* (o÷4)+ω

f. *COS* ω+o2   *SIN* ω+o2
    COS ω   *SIN* ω

Remark. If P is a constant for which:

F ω+P
F ω

and if P is the smallest such constant, then the function F is said to be <u>periodic</u>, and to have <u>period</u> P. That is, as the argument X increases or decreases the values of F repeat themselves after every interval of length P.   In agreement with identity f above, both *SIN* and *COS* are periodic with period o2.

6.57 We have already encountered   (possibly)   periodic functions in Ex 4.29, namely those functions F for which

($\underline{D}F$ ω)*2
1-(F ω)*2

In this exercise through Ex 6.61 we will examine functions for which the above identity is valid. Note   that   it is valid for the functions 1+0×ω and ¯1+0×ω; we will assume that the functions we are analyzing are not these simple ones.

Show that the derivative of the first expression in the above identity is   2×($\underline{D}F$ ω)×$\underline{DD}F$ ω   and that the derivative of the   second expression is -2×(F ω)×$\underline{D}F$ ω. Conclude that, if $\underline{D}F$ ω is not equivalent to 0×ω, then -$\underline{DD}F$ ω equals F ω.

6.58 As a converse to Ex 6.57, show that if -$\underline{DD}F$ ω equals F ω then there is a constant C for which C+1-(F ω)*2 equals ($\underline{D}F$ ω)*2.

6.59 Suppose that the first identity in Ex 6.57 is valid for F and define G:$\underline{D}F$ ω. Show that:

a. $\underline{D}G$ ω equals -F ω and $\underline{D}F$ ω equals G ω;
b. $\underline{DD}G$ ω equals -G ω and $\underline{DD}F$ ω equals -F ω;
c. $\underline{DDD}G$ ω equals F ω and $\underline{DDD}F$ ω equals -G ω;
d. $\underline{DDDD}G$ ω equals G ω and $\underline{DDDD}F$ ω equals F ω;

e. All higher derivatives of F and G equal either F ω or -F ω or G ω or -G ω.   Make a table of the first twenty-five derivatives of F and G.

6.60 Suppose that *F* and *G* are as in Ex 6.59 and suppose also that 0=*F* 0 and 1=*G* 0. Use the table from Ex 6.59e to produce a table of values of the first twenty-five derivatives of *F* and *G* for the argument 0.

6.61 a. Suppose that *F* and *G* are as in Ex 6.60 and use the table in that exercise to produce the coefficient vectors *A* and *B* of the fifteenth order Taylor polynomials of *F* and *G* respectively at 0.

b. For the matrix *T* in Table 6.2.7, evaluate *A POLY T*[;0] and *B POLY T*[;0] and compare the values to those in that table.

6.62 If a table of values of a monadic function is known, approximate values for arguments other than those in the table can be produced by linear interpolation. Suppose that *X* is a vector of arguments and *Y* is the corresponding vector of values of a function *F*. Also, suppose that the elements of *X* are arranged in increasing order, that is, 0<*DIFF X*. Define *M*←(*DIFF Y*)÷*DIFF X*. Then for an argument *A* such that ∧/,(*X*[0]<*A*)∧*A*<*X*[¯1+⍴*X*], an approximate value *V* of *F* can be produced by the sequence *I*←¯1++/*A*∘.≥*X* and *V*←*Y*[*I*]+*M*[*I*]×*A*-*X*[*I*]. Use linear interpolation and Table 6.3.5 to produce approximations to the following values of the circular functions *C* and *S*.

a. *C* 10 c. *C* 82.5
b. *S* 50 d. *S* ¯12

6.63 Table 6.3.7 is a table of values of *COS* and *SIN* on the interval of arguments (0≤⍵)∧⍵≤○0.5. Use the appropriate identities in Ex 6.56 to extend this table to a table of values on the interval of arguments (0≤⍵)∧⍵≤○1.

6.64 Repeat Ex 6.63 with the interval (0≤⍵)∧⍵≤○2 in place of (0≤⍵)∧⍵≤○1.

6.65 Repeat Ex 6.63 with the interval (○2)≥|⍵ in place of (0≤⍵)∧⍵≤○1.

6.66 Use linear interpolation and the tables constructed in Exs 6.63-65 to produce approximations to each of the following values of *COS* and *SIN*.

a. *COS* 2 c. *COS* ¯5
b. *SIN* 4.9 d. *SIN* ¯1.55

6.67 a. Use Table 6.3.7 to sketch the graphs of the functions *COS* and *SIN* on the interval (0≤⍵)∧⍵≤○0.5.

b. Use the sketches in part a and the appropriate identities in Exs 6.56 to sketch the graphs of *COS* and *SIN* on the interval (○5)≥|⍵.

6.68 The following table of values of *COS* and *SIN* was produced by the procedure used in Section 6.3, starting with the first two rows of Table 6.3.7.

```
      4⍕TS
.0000 1.0000  .0000
.0245  .9997  .0245
.0491  .9988  .0491
.0736  .9973  .0736
.0982  .9952  .0980
.1227  .9925  .1224
.1473  .9892  .1467
.1718  .9853  .1710
.1963  .9808  .1951
```

If the scalar *X* is an argument from Table 6.3.7 and *S* is an argument from the above table, then the function values *COS X+S* and *SIN X+S* can be produced using the addition formulas for *COS* and *SIN*. For example, *COS* .5890+.0245 can be produced as follows:

```
      4⍕T[3;]                              4⍕TS[1;]
.5890 .8315 .5556                    .0245 .9997 .0245

      4⍕-/T[3;1 2]×TS[1;1 2]
.8176
```

Use the procedure illustrated above to produce the following values of *COS* and *SIN*.

a. *COS* .7584+.1473
b. *SIN* .3927+.0491
c. *COS* 1.1781-.0736
d. *SIN* .9817+2×.0245

6.69 Use the addition formulas for *SIN* and *COS* to prove the identities in Ex 6.56.

6.70 Use the addition formulas for *SIN* and *COS* to prove identities 6.3.16 and 6.3.17.

6.71 Use the fact that 1=((*COS S*)*2)+*SIN S*)*2 to show that for the functions *DS* and *DC* defined in Section 6.3:

```
      (SIN S)×DS S
      -(1+COS S)×DC S
```

As the scalar *S* is assigned values closer and closer to zero, the values of *SIN S* approach 0 and the values of *COS S* approach 1. Use the above identity to show that -2×*B* equals 0×*A*, where the scalars *A* and *B* are those defined in Section 6.3. Conclude that *B* is 0.

6.72 Show that

a. the second derivatives of the functions 1○ω and 2○ω are -1○ω and -2○ω respectively;

b. the third derivatives of 1○ω and 2○ω are -2○ω and 1○ω respectively;

c. the fourth derivatives of 1○ω and 2○ω are 1○ω and 2○ω respectively;

d. all higher derivatives of 1○ω and 2○ω are either 1○ω or -1○ω or 2○ω or -2○ω.

6.73 a. Show that the function of the positive integer *N* defined by the expression (*N*ρ1 0 ¯1 0)÷!ι*N* produces coefficient vectors of Taylor polynomials of order *N*-1 at 0 of the function 1○ω. Compare the fifteenth order vector with the vector *A* of Ex 6.61a.

b. Define the function of the positive integer *N* that produces coefficient vectors of Taylor polynomials at 0 of the function 2○ω.

c. Repeat part b for a scalar argument *A* in place of 0.

d. Repeat part b for a scalar argument *A* in place of 0 and the function 1○ω in place of 2○ω.

6.74 To devise a practical definition of the growth function *E*, its addition formula is used to define a procedure for producing values of *E* for all arguments in terms of values of *E* for arguments in the interval (0≤ω)∧ω≤1. In this and the following exercise we will define analogous procedures for the sine and cosine functions.

a. Use Ex 6.56 to verify the following sequence of identities:

```
2○ω
2○(○2)|ω
-2○((○2)|ω)-○1
```

Define the functions

```
A:((○2)|ω)-1
B:(ω<-○÷2)-ω>○÷2
C:(A ω)+(○÷2)×B ω
```

and continue the above sequence of identities as follows:

```
-((~|B ω)×2○C ω)+(B ω)×1○C ω
```

Then show that (○÷2)≥|*C X* for every *X*. Consequently, 2○*X* can be defined in terms of values of *COS* and *SIN* for arguments of magnitude less than or equal to ○÷2.

b. Produce a sequence of identities analogous to the one in part a for the function 1○ω in place of 2○ω.

6.75 Theorem 6.1.2 cannot be applied directly to the Taylor polynomials at 0 of the sine and cosine functions because every other term of their coefficient vectors is zero. However, the function called *EST* that was defined in Section 6.1 can be used to estimate the differences between values of the sine and cosine functions and values of their respective Taylor polynomials at 0.

a. For arguments *X* such that *B*≥|*X*, show that:

(*B EST K*)≥|(2○*X*)-((*N*⍴1 0 ¯1 0)÷!⍳*N*) *POLY X*

b. Produce an inequality analogous to the one in part a for the function 1○ω in place of 2○ω.

c. Using the definition of the function called *EP* in Ex 6.39 as a guide, Ex 6.74 and parts a and b above, determine practical definitions of the functions 1○ω and 2○ω which produce 10 decimal digit approximations of the values of these functions.

6.76 A mass of 100 grams is placed on the end of a spring whose stiffness constant *K* is 50. The mass is pulled to a position 0.5 meters below the equilibrium position and released with initial velocity 0. Use 6.3.27 to define the function called *POS* which describes the resulting motion of the mass relative to its equilibrium position. The maximum distance from the equilibrium position achieved by the mass is called the amplitude of the motion. Zetermine the amplitude of the motion in this example.

6.77 Repeat Ex 6.76 for a mass of 200 grams and a spring whose stiffness constant is 50 if the initial position of the mass is 1 meter above the equilibrium position and if the mass passes through the equilibrium position for the first time ○÷4 seconds after it is released.

6.78 a. Show that +/5 6○ω equals *ω.

b. Show that the derivative of the function 5○ω is 6○ω, and that the derivative of the function 6○ω is 5○ω.

c. Make a table of the higher derivatives of 5○ω and 6○ω.

d. Sketch the graphs of the functions 5○ω and 6○ω.

e. Define functions of the positive integer *N* that produce coefficient vectors of Taylor polynomials of order *N*-1 at 0 of the functions 5○ω and 6○ω.

f. Define the Taylor coefficient functions of 5○ω and 6○ω.

g. Show that -/(6 5○ω)*2 equals 1.

h. Sketch the curve of the points (6○*X*),5○*X*.

6.79 Produce an expression that defines the derivative of each of the following functions.

a. +/3 ¯4×1 2○ω

b. -/5 6○5 ¯10×α

c. (*1+2×α)+6○¯4+α

d. (2○-ω)-3×*5-ω

6.80 Sketch the graphs of each of the following functions.

a. 4×1○2×ω

b. 4×1○4×ω

c. 2×2○6×α

d. 2×2○8×α

e. +/2×1 2○3×ω

6.81 a. Repeat the example in the text for each choice of *C* as 0.5, 1, 1.5, and 2. That is, for each *C* let *S*←0.5 and *TBL*←1 2⍴1 ¯0.6. Then use the functions *POS* and *VEL* to extend this table as in the text. Finally, plot the coordinate points represented by the rows of *TBL*. Be sure to plot enough points to be certain of the behavior of the graph.

b. Discuss the variations in the graphs of part a.

6.82 Suppose that an external force imparts an acceleration of 2○*P*×*T* on the bob at time *T*, where *P* is a constant. Then in place of the acceleration function defined by 6.5.1 we have:

EXA:2○*P*×ω
ACC:(*EXA* ω)-(*C*×*VEL* ω)+*POS* ω

Define the derivatives *DACC* and *DDACC* and repeat Ex 6.81 for the pairs of scalars *C* and *P* below.

C←0.1,*P*←0.8
C←1,*P*←0.8
C←2,*P*←0.8

C←0.1,*P*←1
C←1,*P*←1
C←2,*P*←1

6.83 As another example of producing approximate solutions of initial value problems, consider a bi-molecular reaction, in which two reactants, whose concentrations at time *T* are the function values *F T* and *G T*, combine to produce a substance whose concentration at time *T* is the function value *N T*. According to the basic laws of reaction kinetics, the function DN ω is proportional to both F ω and G ω. That is, there is a scalar K for which K×(F ω)×G ω equals DN ω. Evidently, since F 0 and G 0 are the initial concentrations of the reactants, F ω equals (F 0)-N ω and G ω equals (G 0)-N ω. Therefore, setting A←F 0 and B←G 0, we see that

```
DN ω
K×(A-N ω)×B-N ω.
```

For this exercise we will assign K←1 and A←3 and B←7.

a. Using the above identity as a guide, define DN:(3-N ω)×7-N ω and show that the second derivative of N can be defined as DDN:((-DN ω)×7-N ω)+(3-N ω)×-DN ω. Note that just as DN is defined in terms of N, DDN is defined in terms of N and DN. Define the third derivative of N in terms of N, DN, and DDN.

b. Using the example in the text as a guide and the fact that the initial concentration N 0 is 0, produce approximate values of N 0.2×⍳16 and sketch the graph of these approximate values. (The function N can be defined explicitly in terms of ⋆ω; see Ex 8.32.)

7

7.1 For each value of Y below sketch the graph of the function Y-ω⋆3. Starting with the value Y, geometrically trace Newton's method to show that this rootfinder will produce approximations to the root of this function.

a. Y←4 b. Y←¯5 c. Y←0

7.2 Repeat Ex 7.1 replacing the starting values Y by -Y.

7.3 Another strategy for choosing starting values for Newton's method in order to produce the root of the function in Ex 7.1 is to choose the starting value, to be called S, to be the same for all choices of Y. For example, if S←1 then the function called IF in the text would be defined as

```
IF:ω RTF 1
```

For which S below will this choice of the starting value lead to the root of the function Y-α⋆3 for all scalars Y?

a. S←1 b. S←¯3 c. S←0

7.4 It is worth repeating the development in Section 7.1 for another example. Consider the functions F:α⋆4 and

`G:ω*÷4`. Show that *F* and *G* are inverse functions on the interval `ω≥0`. That is, verify identities 7.1.1.

7.5 Sketch the graph of the function *F* defined in Ex 7.4 on the interval `ω≥0`. Using the example in the text as a guide, produce the graph of the function *G* of Ex 7.4 from the graph of *F*.

7.6 Using the example in the text as a guide, produce a practical definition of the function *G* in Ex 7.4. Test your definition by testing the Identities 7.1.1.

7.7 Using Identity 7.1.4 to show that the derivative of the function `ω*÷4` is the function `(÷4)×ω*-3÷4`.

7.8 Argue for the fact that since `*ω` is a strictly monotonic function whose graph has no inflection points, any choice of the right argument of the function `RTE` within the definition of the function `IE` in the text will produce approximate values of the roots of the function `Y-*α`.

7.9 The proof of Identity 7.2.3 is as follows. If the scalars *Y* and *Z* are members of the domain of `IE` then they are also members of the range of the function `*α`. That is, there exist scalars *U* and *V* for which `Y=*U` and `Z=*V`. Consequently, since `IE` is the inverse function of `*X`, `U=IE Y` and `V=IE Z`. Now verify the following sequence of identities:

```
IE Y×Z
IE (*U)××V
IE *U+V
U+V
(IE Y)+IE Z
```

7.10 Since `1=*0`, it follows that `0=⍟1`. Verify the following sequence of identities and then prove 7.2.7:

```
0
⍟Y÷Y
⍟Y×÷Y
(⍟Y)+⍟÷Y
```

7.11 Prove Identity 7.2.8.

7.12 Determine the derivative and sketch the graph of each of the following functions.

a.	`⍟4×ω`	e.	`*⍟×ω`
b.	`20⍟ω`	f.	`⍟⍟α`
c.	`(X*2)×⍟1+2×ω`	g.	`α÷⍟α*4`
d.	`÷⍟ω`	h.	`(⍟ω)*3`

7.13 Show that the derivative of the function ⍟-ω is the function ÷ω. Then show that the derivative of ⍟|ω is ÷ω.

7.14 Determine the derivative and sketch the graph of each of the following functions.

a. ⍟|¯7×α   b. α×⍟⌊α*3   c. 10⍟|ω

7.15 We will now study another procedure for evaluating the function ⍟ω. The first step is to examine the polynomials defined by 7.2.10 in order to determine the number of terms required to produced approximate values of specified accuracy. As before, we will use Theorem 6.1.2.

If we ignore the term ⍟*A* in 7.2.10 for the moment and replace the argument (ω-*A*)÷*A* with, say, α, the result is the polynomials defined by the expression

```
(0,(¯1*⍳N)÷1+⍳N) POLY α                                  [1]
```

Test the following identity for positive integers *K* and *N* such that *K*<*N*:

```
⌈/QT K↓|(¯1*⍳N)÷1+⍳N
(N-1)÷N
```

Consequently, for the scalar *M* defined in Theorem 6.1.2:

```
M
B×(N-1)÷N
```

Since for large *N*, the scalar defined by the expression (*N*-1)÷*N* is less than 1 but approximately equal to 1, we can instead choose *M* to be equal to *B* (instead of a little smaller than *B*). That is, define *M*←*B*. Then the right hand side of the inequality in Theorem 6.1.2 is

```
((÷K+1)×B*K)÷1-B                                         [2]
```

Note that if *B*<1 then by choosing the slightly larger value for *M* we have increased slightly the value of the right hand side of the inequality in Theorem 6.1.2, which means that we have produced a slightly larger estimate of the values of the suffixes of the polynomials in question.

According to Theorem 6.1.2, the estimate provided by Expression 2 of suffixes of polynomials defined by Expression 1 is valid for arguments *T* such that *B*>|*T* if 1>*M*, or in this case if 1>*B*. Substituting the expression (*X*-*A*)÷*A* for *T*, we see that the estimate is valid for scalars *X* and *A* for which *B*>|(*X*-*A*)÷*A*, or equivalently (*A*×*B*)>|*X*-*A*, if 1>*B*. Consequently the Taylor polynomials at *A* for ⍟ω defined by 7.2.10 are approximations of ⍟ω for arguments in the interval *A*>|ω-*A*.

7.16 Use Expression 2 of Ex 7.15 to define a function which is analogous to the function *EST* defined in Section 6.1. Some experimentation with this function will show that a very large number of terms of the Taylor polynomials of ⍟ω at *A* are needed to produce reasonable approximations to function values ⍟*X* unless *X* is quite close to *A*. After experimenting with this function, show that the values of the expression

```
(⍟A)+(0,(¯1*⍳9)÷1+⍳9) POLY (X-A)÷A
```

are 10 decimal digit approximations to ⍟*X* if 0.1>|(*X*-*A*)÷*A*.

7.17 Use Ex 7.16 to show that the following function will produce 10 decimal digit approximations of ⍟*X* for arguments *X* in the interval (0≤ω)∧ω≤10.

```
A←1+.1×⍳90
VA←⍟A
AL:VA[I]+(0,¯1*⍳9) POLY (ω-A[I])÷A[I←⌊10×ω-1]
```

7.18 Use the multiplication rule for ⍟ω (Identity 7.2.6) to extend the domain of the function in Ex 7.17 to approximate values of ⍟ω on the interval (10≤ω)∧ω≤100. Do not use values of the primary function ⍟ω other than those in the vector *VA* of Ex 7.17.
Hint. ⍟ω and ⍟10×0.1×ω and (⍟10)+⍟0.1×ω are equivalent, and if (10≤ω)∧ω≤100, then (1≤0.1×ω)∧10≥0.1×ω.

7.19 Extend the domain of the function in Ex 7.17 to produce approximate values of ⍟*X* for all arguments *X* such that *X*>0. Do not use values of the primary function ⍟*X* other than those in the vector *VA* of Ex 7.17.

7.20 The approximation procedure described in Exs 7.17-19 uses values of the primary function ⍟ω for the vector of arguments *A*←1+.1×⍳90. It is not difficult to produce 10 decimal digit approximations to the values ⍟*A*, thus making the approximation procedure independent of the primary function ⍟ω. A vector of approximately 10 decimal digit approximations to ⍟*A*, which will be denoted by *VA*1, can be produced as follows:

```
VA1←SUM (0,(¯1*⍳9)÷1+⍳9) POLY (DIFF A)÷¯1↓A
```

Test the fact that the elements of *VA*1 are approximately 10 digit approximations to the corresponding elements of the vector *VA* in Ex 7.17.

7.21 Exercises 7.17-7.20 provide another practical definition of the logarithm function ⍟ω. Once the function ⍟ω has been defined it can be used in conjunction with Newton's method to produce approximate values of its inverse function, namely *α. Use the logarithm function ⍟ω and Newton's method to produce approximate values of *¯5+⍳11.

7.22 Show that the addition formula 7.2.16 is valid for the function *B*⋆⍵ defined by Identity 7.2.13.

7.23 Use 7.2.13 and 7.2.14 to show that for a scalar *B* for which (*B*>0)∧1≠*B*, the function denoted by *B*⋆⍺ and the function denoted by *B*⍟⍵ are inverses.

7.24 Produce an expression that defines the derivative of each of the following functions.
Hint. For functions involving the dyadic logarithm and power functions use the defining Identities 7.2.13 and 7.2.14.

a.	⍟⍵⋆2	e.	⍵⍟⍵⋆3
b.	2⋆⍵	f.	⍟2○⍺
c.	⍺⋆⍺	g.	(1○⍵)⋆2○⍵
d.	5.5⍟⍺	h.	⍵⋆⋆⍵
		j.	⍟|⍵

7.25 Use 7.2.13 to show that for scalars *A*, *B* and *C*:

(*A*⋆*B*)⋆*C*
A⋆*B*×*C*

Then use this identity to show for a positive integer *N* that the function ⍺⋆*N* and the function ⍵⋆÷*N* are inverse.

7.26 Sketch the graph of the inverse function of the function ⍵⋆*N* for each integer *N* from 1 to 5.

7.27 For each integer *N* from 1 to 5 and each argument *Y* of the inverse function of the function *F*:⍵⋆*N*, devise a strategy for choosing the right argument *S* of the function *R̲TF* so that *Y* *R̲TF* *S* is a vector of approximate values of *Y*⋆÷*N*. Precisely, define *S* as the value of a dyadic function whose arguments represent *Y* and *N*.

7.28 If *N* is an even positive integer then the function ⍺⋆*N* is a decreasing function on the interval 0≥⍺, and consequently has an inverse function on this interval. Define this inverse function in terms of the function ⍵⋆÷*N*.

7.29 For a scalar *S*, show that *QT* *S*⋆1+⍳*N* equals (*N*-1)⍴*S*, while *QT* (1+⍳*N*)⋆*S* equals (*QT* 1+⍳*N*)⋆*S*.

7.30 a. Sketch the graph of the function *ARCSIN*.

b. Define the function *R̲TC*1 for *C*1:1○⍵. Then define a monadic function *START* whose domain is the domain of *ARCSIN* and whose value *S*←*START* *Y* is a right argument of the function *RTC*1 such that *Y RTC*1 *S* is an approximate value of *ARCSIN Y*.

c. Since 0=*SIN* 0 it follows that 0=*ARCSIN* 0. Also, *DARCSIN*:÷0○ω. Consequently, according to the First Fundamental Theorem of the Calculus, a practical definition of *ARCSIN* is 0 *SDARCSIN* ω, where *SDARCSIN* is the approximate definite integral function defined as in Section 5.5. Use the graph in part a to determine those arguments of ÷0○ω near which this function is not like a polynomial. (Hint: where does ÷0○ω have a vertical tangent? See Ex 5.81.) Test how well this practical definition of *ARCSIN* works near these arguments.

7.31 In sections 7.2 and 7.3 we were able to recognize patterns in the higher derivatives of the functions ⍟ω and ω⋆*R* and this enabled us to produce expressions defining coefficient vectors of Taylor polynomials of any order for these functions. Discovering a pattern in the higher derivatives of the function *ARCSIN* is not so easy. Starting with Identity 7.4.1, produce expressions defining the first five derivatives of the function *ARCSIN*, and search for a pattern from which you can define coefficient vectors of Taylor polynomials of any order for the function *ARCSIN*.

7.32 There is another approach to defining the coefficient vectors of Taylor polynomials of the function *ARCSIN*. According to Identity 7.4.1, the function *ARCSIN* is an integral of the function ÷0○ω.

a. For a scalar *A* and an integer *N* the vector *N TC*0 *A* is the coefficient vector of the *N*th order Taylor polynomial of 0○ω at *A*. Show that the vector defined by the expression (*ARCSIN A*) *INTEG N TREC N TC*0 *A* is the coefficient vector of the *N*th order Taylor polynomial of *ARCSIN* at *A*. Define the Taylor coefficient function *TCN*1 of ¯1○ω.

b. The coefficient vectors defined in part a are useful in that they provide lists of the higher derivatives of the function *ARCSIN*. However, just as with the Taylor polynomials of ⍟ω and ω⋆*R* the Taylor polynomials of *ARCSIN* cannot be used directly to produce approximate values of *ARCSIN* since values of *ARCSIN* appear in the coefficient vectors. To approximate values of *ARCSIN* with polynomials we can use a procedure similar to the one used for ⍟ω in Ex 7.20. Suppose that the integer *N* is such that for every scalar *A* for which 0.9≥|*A*, the polynomial with coefficient vector defined in part a is a 10 decimal digit approximation to values of *ARCSIN* on the interval 0.1>|*Y*-*A*. Define a procedure similar to that in Ex 7.20 to produce (approximately) 10 decimal digit approximations of values of *ARCSIN* on the interval 1≥|ω.

c. The question to be considered in this part is how to determine the integer *N* described in part b. As before, we could appeal to Theorem 6.1.2. However, as the expressions defining coefficient vectors of Taylor polynomials become more and more complicated it becomes more difficult to apply that theorem, and a practical procedure must be found. For the coefficient vector *C* of a Taylor polynomial of a monadic scalar function *F* at an argument *A*, a good rule of thumb is to accept the values *C POLY X-A* as *K* digit approximations of *F X* if the values *C POLY X-A* and (¯1↓*C*) *POLY X-A* agree to *K* digits. Use this rule to determine the integer *N* in part a. Note that this procedure can be used to estimate the errors in the approximation method discussed in Section 6.5.

7.33 The cosine function 2○ω is a decreasing function on the interval (0≤ω)∧ω≤○1 and so has an inverse function on that interval, which we will call *ARCCOS*. Repeat Ex 7.30 for the function *ARCCOS* in place of *ARCSIN*.

7.34 a. Devise a variation of the argument in Section 7.4 to show that the derivative *DARCCOS* ω equals -÷0○ω. Then use the fact that both *ARCSIN* ω and -*ARCCOS* ω are integrals of ÷0○ω to show that *ARCCOS* ω equals (○÷2)-*ARCSIN* ω.

b. Define the Taylor coefficient function *TCN*2 of ¯2○ω.

7.35 The cosine function 2○ω is strictly monotonic on each of the intervals (ω≥○*M*)∧ω≤○*M*+1, where *M* is an integer, and therefore has an inverse function on each of these intervals. Define these inverse functions in terms of the function ¯2○ω.

7.36 The tangent function denoted by 3○ω is defined by (1○ω)÷2○ω. Show that the derivative of 3○ω is ÷(2○ω)*2. Show also that ÷(2○ω)*2 is 1+(3○ω)*2. The tangent function is increasing on the interval (○÷2)≥|ω and therefore has an inverse function on this interval, which we call *ARCTAN*. Use Identity 7.1.4 to show that *DARCTAN* ω equals ÷1+ω*2.

7.37 Repeat Ex 7.30 for the function *ARCTAN* in place of the function *ARCSIN*.

7.38 Repeat Ex 7.31 for the function *ARCTAN* in place of the function *ARCSIN*.

7.39 We will now repeat Ex 7.32 for the function *ARCTAN* in place of *ARCSIN*. According to Ex 7.36, the function *ARCTAN* is an integral of the function ÷1+ω*2.

a. Repeat 7.32a by first defining an expression which produces coefficient vectors of Taylor polynomials of the function *ARCTAN*.

b. Repeat 7.32b.
c. Repeat 7.32c.

7.40 The tangent function 3○ω is strictly monotonic on each of the intervals (ω≥○*M*-0.5)∧ω≤○*M*+0.5, and therefore has an inverse function on each of these intervals. Define these inverse functions in terms of the function ⁻3○ω.

7.41 Produce an expression that defines the derivative of each of the following functions.

a. ¯3○2+4×ω   c. α×¯1○α
b. (¯1○ω)×*ω*2   d. (¯1○α)+¯2○3×α

7.42 Show that *D̲ARSINH* ω equals ÷4○ω and then repeat Exercise 7.30 for *ARSINH* in place of the function *ARCSIN*.

7.43 Repeat Exs 7.31 and 7.32 for the function ¯5○ω in place of ¯1○ω.

7.44 The hyperbolic cosine function denoted by 6○ω is an increasing function on the interval ω≥0 and therefore has an inverse function on this interval, which we call *ARCOSH*. Show that *D̲ARCOSH* ω equals ÷¯4○ω and then repeat Ex 7.30 for the function *ARCOSH* in place of *ARCSIN*.

7.45 Modify the argument in Section 7.5 to show that *ARCOSH* ω equals ⍟ω+¯4○ω.

7.46 The hyperbolic cosine function is a decreasing function on the interval ω≤0 and therefore also has an inverse function on this interval. Define this inverse function in terms of the function *ARCOSH*.

7.47 Repeat Exs 7.31-7.32 for the function ¯6○ω, or equivalently *ARCOSH* ω, in place of ¯1○ω.

7.z8 The <u>hyperbolic tangent</u> function denoted by 7○ω is defined by (5○ω)÷6○ω. Show that the derivative of 7○ω is ÷(6○ω)*2. The hyperbolic tangent function is increasing on the interval of all scalars and therefore has an inverse function on this interval, which we call *ARTANH*. Show that *D̲ARTANH* ω equals ÷1-ω*2 and then repeat Ex 7.30 for the function *ARTANH* in place of *ARCSIN*.

7.49 In this exercise we will produce an explicit rule for the function *ARTANH* similar to those for *ARSINH* and *ARCOSH*. To construct this rule we must produce an expression whose value for an argument *Y* such that 1>|*Y* is identical to the solution *X* of the equation *Y*=6○*X*, or equivalently *Y*=((**X*)-*-*X*)÷(**X*)+*-*X*. Show that if we let *U*←**X* then *Y*=(¯1+*U**2)÷1+*U**2 and equivalently *U*=((1+*Y*)÷1-*Y*)*0.5.

Consequently, *X*=0.5×⍟(1+*Y*)÷1-*Y*. That is:

```
ARTANH Y
0.5×⍟(1+Y)÷1-Y
```

7.50 Repeat Exs 7.31-32 for the function ¯7○ω, or equivalently *ARTANH* ω, in place of the function ¯1○ω.

7.51 Produce an expression which defines the derivative of each of the following functions.

```
a.  ¯5○3+¯2×ω              d.  ¯4○*α
b.  (¯1○ω)+¯6○ω            e.  ¯2○1○α
c.  (*α)×¯3○α*2            f.  (¯7○ω*0.5)-¯6○ω
```

7.52 The polynomials (0,(¯1*⍳*N*)÷1+⍳*N*) *POLY* ω-1 are, according to Ex 7.15, approximations of ⍟ω on the interval 1>|ω-1. We will now examine the values of these polynomials at the endpoints 0 and 2 of this interval. For the argument 0 we have:

```
(0,(¯1*⍳N)÷1+⍳N) POLY 0-1
+/(0,(¯1*⍳N)÷1+⍳N)×¯1*⍳N+1
¯1×+/0,÷1+⍳N
-+/÷1+⍳N
```

and for the argument 2:

```
(0,(¯1*⍳N)÷1+⍳N) POLY 2-1
+/(¯1*⍳N)÷1+⍳N
-/÷1+⍳N
```

a. Consider the last expression in the first sequence of identities. The values of the expression +/÷1+⍳*N* grow without bound as *N* is assigned large values. To see that this is true consider the function *P*:+/(2*ω-1)÷1+⍳2*ω. *P* 1 is +/÷2, *P* 2 is +/÷3 4, *P* 3 is +/÷5 6 7 8, and so on. It is not difficult to see that *P K* is greater than +/(2**K*-1)⍴2**K*, or equivalently (2**K*-1)×÷2**K*, or ÷2. Therefore, for *S*:(*S* ω-1)+*P* ω:ω=0:1 we have that *S K* is greater than +/*K*⍴÷2, or equivalently *K*÷2. Also, *S K* is identical to +/÷1+⍳2**K*, and so (+/1+⍳2**K*)>*K*÷2. Finally, for *R*:⌊2⍟ω we have that *N*>2**R N* and consequently (+/÷1+⍳*N*)>(*R N*)÷2. Now argue for the fact that the values of +/÷1+⍳*N* grow without bound as *N* is assigned large values.

b. Consider the last expression in the second sequence of identities. The values of the expression -/÷1+⍳*N* approach a unique value as *N* is assigned large values. To see that this is true we must estimate the values of the suffices -/*J*↓÷1+⍳*N*. It is not difficult to check that since the elements of the vector ÷1+⍳*N* are positive and in decreasing order, |-/*J*↓÷1+⍳*N* is smaller than 1↑*J*↓÷1+⍳*N*, or equivalently ÷*J*+1. Therefore:

```
(÷1+N⌊M)>|(-/÷1+⍳N)--/÷1+⍳M
```

For example, if both *N* and *M* are greater than 1000 then -/÷1+⍳*N* and -/÷1+⍳*M* differ by at most .001. Now argue as in Section 6.1 to show that the values of the expression -/÷1+⍳*N* approach some other value as *N* is assigned large values. Although we will not prove it here, the value approached by -/÷1+⍳*N* as *N* is assigned large values is ⍟2.

7.53 Define the function *C* whose value *C N* is the coefficient vector of the *N*th order Taylor polynomial at 0 of the function ¯3○*ω* (see Ex 7.39a). These polynomials are approximations to ¯3○*ω* on the interval 1>|*ω*. We will now examine the values of these polynomials at the endpoint 1 of this interval. Show that

```
(C N) POLY 1
-/÷1+2×⍳⌊0.5×N+1
```

Using Ex 7.52b as a guide, show that the values -/÷1+2×⍳⌊0.5×*N*+1 approach a unique value as *N* is assigned large values. Although we will not prove it here, that value is ¯3○1, or ○÷4.

7.54 a. The *N*th power of the square matrix *A* is defined to be the matrix product of *N* copies of *A*. For example, the second power or square of *A* is *A*+.×*A*, the third power or cube of *A* is *A*+.×*A*+.×*A*, and so on. In this regard *ID* 1↑⍴*A* is called the zeroth power of *A*. Compute the second through fifth power of each of the following matrices.

```
       A                          B
  12   24   54             ¯1      ¯12
 ¯15  ¯32  ¯75             1.5       8
   4    9   22

       C
 9.5      3      14      0
 3.75     9.5     3     14
¯3.5      0      ¯8      3
¯0.75    ¯3.5     0     ¯8
```

b. Suppose that *AN* is the *N*th power of *A* and *AN*1 is the (*N*-1)th power. Use the associativity of the matrix product to show that if *A* is a valid right argument of domino:

```
∧/,AN1=AN+.×⌹A
```

Thus ⌹*A* is called the ¯1th power of *A*, (⌹*A*)+.×⌹*A* the ¯2th power of *A*, and so on.

7.55 a. Polynomials can be evaluated using only `+` and `×`. For example, show that `5 4 ¯7 2 POLY ω` equals `5+ω×4+ω×¯7+ω×2`. More generally, for

```
POL:α[0]+ω×(1↓α) POL ω : 0=ρα : 0
```

show that `C POL ω` equals `C POLY ω` for a vector `C`.

b. Replacing the function `×` with `+.×` in the definition of `POL` produces the matrix product polynomial function

```
MPOLY:(α[0]×ID 1↑ρω)+ω+.×(1↓α) MPOLY ω:0=ρα:(ρω)ρ0
```

In particular, `(I=⍳N) MPOLY A` is the *I*th power of the matrix *A*. Test this fact for various choices of *I* and *N* and the matrices in Ex 7.54a.

7.56 a. Recall from Ex 3.35 that for a square matrix *A* the function `DET A-ω×ID 1↑ρA` is a polynomial of degree `1↑ρA` called the characteristic polynomial of *A*. The coefficient vector *CA* of this polynomial can be produced by the method of fitting polynomials (Ex 3.57). Determine the coefficient vector of the characteristic polynomial of each matrix in Ex 7.54a.

b. The Cayley-Hamilton Theorem states that if *CA* is the coefficient vector of the characteristic polynomial of a square matrix *A* then `∧/∧/0=CA MPOLY A`. Test this theorem for each matrix in Ex 7.54a. The proof is an advanced topic in courses on linear algebra.

c. The roots of the characteristic polynomial of a square matrix *A* are called the eigenvalues or singular values of *A*. Use Newton's method to produce the eigenvalues of each matrix in Ex 7.54a.

d. The matrix `A-R×ID 1↑ρA` is a valid right argument of domino if and only if the scalar *R* is not an eigenvalue of *A*. Use this fact to show that if *CA* is the coefficient vector of the characteristic polynomial of *A*, then *A* is a valid right argument of domino if and only if `0≠CA[0]`.

7.57 a. Suppose that *CA* is the coefficient vector of the characteristic polynomial of *A* and that *A* is a valid right argument of domino. Verify the sequence of identities:

```
(ρA)ρ0
CA MPOLY A
(CA MPOLY A)+.×⌹A
(CA[0]×⌹A)+(1↓CA) MPOLY A
```

and consequently:

```
⌹A
-(1↓CA÷CA[0]) MPOLY A
```

b. The last identity in part a can be used to produce the inverse matrix ⌹*A*. For example, ¯216 354 80 ¯1 *POLY* ⍵ is the characteristic polynomial of 3 3⍴(⍳9)*2:

```
      3⍕-(1↓¯216 354 80 ¯1÷¯216) MPOLY 3 3⍴(⍳9)*2
  .931  ¯.611   .181
¯1.500   .667  ¯.167
  .625  ¯.167   .042

      3⍕⌹3 3⍴(⍳9)*2
  .931  ¯.611   .181
¯1.500   .667  ¯.167
  .625  ¯.167   .042
```

Use this procedure to produce the inverse of each matrix in Ex 7.54a.

c. The last identity in part a can be used as a definition of ⌹⍵ if we can produce coefficient vectors of characteristic polynomials without using domino (the definition of *CFP* is based on domino). This can in fact be done by using the trace function *TR*:+/0 0⍉⍵ which produces the sum of the diagonal elements of its arguments. For example, ¯216 354 80 ¯1 is the coefficient vector of the characteristic polynomial of 3 3⍴(⍳9)*2, and this vector can be produced as follows:

```
      M←3 3⍴(⍳9)*2
      T1←TR M
      T2←TR M+.×M
      T3←TR M+.×M+.×M
      C←T1,¯1
      C←(-(C+.×T1,T2)÷2),C
      C←(-(C+.×T1,T2,T3)÷3),C
      C
¯216 354 80 ¯1
```

Use this procedure to produce the coefficient vector of the characteristic polynomial of each matrix in Ex 7.54a.

8

8.1 Evaluate each of the following:

```
a.  1 0 1 0 1/'ABCDE'            d.  ⍎'3+4*2'
b.  ((⍳4)∘.+⍳6)∊2×⍳5             e.  4 6⍴6↑'XYZ'
c.  (10⍴'ABC')⍳'BCD'             f.  '4 5 6'∘.=4 5 6
```

8.2 Suppose that *C̲F* and *C̲G* are the cline functions of *F* and *G* respectively and suppose that *X* is an argument for which (0=*F X*)∧0=*G X*. Then

```
F X+S            G X+S            (F X+S)÷G X+S
S×S C̲F X         S×S C̲G X         (S C̲F X)÷S C̲G X
```

As *S* is assigned values close to 0 the values (*S* C̲*F X*)÷*S* C̲*G X* approach (0 C̲*F* *X*)÷0 C̲*G X*, or equivalently (D̲*F X*)÷D̲*G X*. This value is also approached by the values of (D̲*F X*+*S*)÷D̲*G X*+*S*. Thus, according to the preceding identity, we have:

<u>L'Hospital's Rule</u>. If *X* is an argument for which (0=*F X*)∧0=*G X* then as *S* is assigned values close to 0 the values approached by the values (*F X*+*S*)÷*G X*+*S* and (D̲*F X*+*S*)÷D̲*G X*+*S* are identical.

For example, the function *H*:(1○⍵)÷⍵ is equivalent to (*F* ⍵)÷*G* ⍵, where *F*:1○⍵ and *G*:⍵. Evidently, (0=*F* 0)∧0=*G* 0. Since D̲*F*:2○⍵ and D̲*G*:1+0×⍵, the values (D̲*F S*)÷D̲*G S*, or equivalently 2○*S*, approach 1 as *S* is assigned values close to 0. Therefore the values (*F S*)÷*G S*, or equivalently *H S*, also approach 1 as *S* is assigned values close to 0.

Show that the values of the function (1-2○⍵)÷⍵ approach 0 for arguments near 0.

8.3 For each function and scalar *A* below, use L'Hospital's rule to determine the value approached by the function values for arguments near *A*.

```
a.  (¯1+*⍺)÷⍺     ;A←0        c.  (¯9+⍵*2)÷¯3+⍵  ;A←3
b.  (3○⍺)÷⍺       ;A←0        d.  (⍵-1○⍵)÷⍵*3    ;A←0
```

Hint: apply L'Hospital's rule more than once in part d.

8.4 If a sphere is expanding at the rate of 50 cubic feet per minute, at what rate is its radius expanding when the radius is equal to 10 feet?
Hint. This problem is known as a <u>related rates</u> problem, the point being that the problem requires you to establish an identity relating the rate of change in volume to the rate of change in the radius. Suppose that the values of the functions called *VOL* and *RAD* are the volume and radius of the sphere at time *T* respectively. Then:

```
VOL T
○(4÷3)×(RAD T)*3
```

Use the Composition Rule for Derivatives to show for the derivatives D̲*VOL* and D̲*RAD* of *VOL* and *RAD* respectively:

```
D̲VOL T
○4×((RAD T)*2)×D̲VOL T
```

Now solve the above related rates problem. Note that we do not know expressions to define *RAD* and *VOL*.

8.5 Suppose that a 25 foot ladder is leaning against a wall and its foot is pulled away from the wall at the rate of 4 feet per second. How fast is the height of the tip of the

ladder decreasing when the foot is 10 feet from the base of the wall? 20 feet? 25 feet?
Hint. Suppose that *HEIGHT* and *BASE* are functions such that at time *T*, *HEIGHT T* is the height of the tip of the ladder against the wall and *BASE T* is the distance from the base of the wall to the tip on the ground. Then ((*HEIGHT T*⋆2)+(*BASE T*)⋆2 equals the constant 25⋆2. Now use the Composition Rule for Derivatives to solve this related rates problem.

8.6 For scalars *X* and *Y*, the set of all coordinate points *X*,*Y* for which (*X*⋆2)+2×*Y*⋆2 equals 1 is an ellipse. If a particle is moving around this ellipse then there are monadic scalar functions *XPOS* and *YPOS* for which ((*XPOS T*)⋆2)+2×(*YPOS T*)⋆2 equals 1 and the point (*XPOS T*),*YPOS T* is the position of the particle at time *T*. Suppose that *DXPOS T* equals *YPOS T* and define the function *DYPOS* in terms of the functions *XPOS* and *YPOS*. Is the particle moving in the clockwise or counterclockwise direction?

8.7 In order to review the definition of the function *D* of Section 8.2 it is helpful to define the monadic function *T*, which is analgous to *D* but produces Taylor coefficient functions. For example, the value *T* '(2○ω)×⋆ω', obtained by applying the Times Rule for Taylor coefficient functions, is

(α *TC*2 ω) *TTIMES* α *TE* ω

Define *T*, using the definition of *D* as a guide. Then define the dyadic plus rule function *T*1, the monadic minus rule function *T*2, and the dyadic minus rule function *T*3. The remaining Taylor coefficient function rules should be defined when the corresponding derivative rules are defined in subsequent exercises.

8.8 a. Define the dyadic times rule function *D*5 and the monadic divide rule (reciprocal rule) function *D*6.

b. Use the functions *D*5 and *D*6 to define the dyadic divide rule function *D*7.

c. Define the monadic power rule function *D*8. Be sure that your function applies to expressions of the form '⋆*F* ω', and not just '⋆ω' and '⋆α'.

d. Define the monadic logarithm rule function *D*10.

e. Define the dyadic power rule function *D*9 and the dyadic logarithm rule function *D*11.

f. Recall that the monadic function ○ω is *PI*×ω. Define the monadic circle rule function *D*12.

g. The dyadic circle function α○ω is a family of monadic functions, ranging from the tanh function 7○ω to its inverse function ¯7○ω. For convenience we will name these functions *C7* through *CN7*. That is, *C7*:7○ω and *CN7*:¯7○ω. Also, *C1*:1○ω, and so on. Their derivatives are then named *DC7* through *DCN7*. We will name the corresponding derivative rule functions *DRC7* through *DRCN7*. For example, the derivative rule function for the sine function is defined as follows:

```
RC1←'2○'
DRC1:(ENP RC1,ω),PFS[2],D ω
```

Define the remaining derivative rule functions for the dyadic circle functions. The dyadic circle function rule *D13* can then be defined by:

```
NN←'N'
RC←'DRC'
CFN:RC,((0>⍎α)/NN),⍕|⍎α
D13:⍎(CFN α),BL,ENQ ω
```

8.9 Define an integral of each of the following functions. Hint. The plus and scale rules should be applied formally until you are familiar with them. If the plus rule is to be applied to a function *H*, first define the summands *F* and *G*, then define integrals of *F* and *G*, and finally apply the plus rule. If the scale rule is to be applied to a function *G*, first define the function *F* and scalars *A* and *B* for which *G* ω is *F A+B*×ω, then define an integral of *F*, and finally apply the scale rule.

```
a.  *¯4+3×ω
b.  (2×1○2×α)+4×2○α
c.  ÷4○1-0.5×ω
d.  -5○○α
e.  ÷1-α÷3
f.  4*2×ω
g.  +/*5 ¯3 2×α
h.  -/2 ¯2×5 6○ω
i.  ÷0○α+α
j.  3÷1+ω*2
k.  1+4×3○α÷4
l.  +/5 6×2 3○ω
m.  2÷4+ω*2
n.  (1+4×α)*4
o.  3*¯1+¯5×ω
p.  7○ω÷2
```

8.10 Reproduce the table of integrals in Section 8.3 using the names *LN* α for ⍟α, *COS* α for 2○α, *ARTANH* α for ¯7○α, etc.

8.11 Define an integral of each of the following functions.

```
a.  COS ¯3+ω
b.  3×E 5-4×α
c.  (2×SINH 3×ω)-4×COSH ¯5×ω
d.  -TAN 3+α
e.  2×ARTANH 5+ω
f.  ARSINH α÷4
g.  3×SIN ¯4+3×ω
h.  4×LN ω*2
i.  ¯2+4×ARCCOS 1+3×α
j.  TAN 1-ω
```

8.12 Define an integral of each of the following functions. Hint. These exercises illustrate the last integral rule of Section 8.3. As with the plus and scale rules in Ex 8.9, this rule should be applied formally until you are familiar with it.

```
a.  2×(¯6+2×α)*0.5
b.  (3×ω*2)÷(1+ω*3)*2
c.  α×SIN α*2
d.  (2○α)×(1○α)*2
e.  (*ω)÷1+*ω
f.  (E (1+α)*0.5)÷(1+α)*0.5
g.  (5○ω)÷(6○ω)*÷3
h.  (SIN α)÷3+COS α
i.  (7○ω)×(6○ω)*2
j.  (ARCTAN ω)÷1+ω*2
k.  (¯5+¯1○α)÷0○α
l.  ¯5÷4○5×ω
m.  (¯6○○α)÷¯4○○α
n.  (α+LN 2+α*2)÷2+α*2
```

8.13 Use the addition formulas for the sine and cosine functions (Section 6.3) to show that

```
(1○ω)*2                (2○ω)*2
0.5×1-2○2×ω            0.5×1+2○2×ω
```

8.14 Use the identities in Ex 8.13 to produce an expression that defines an integral of each of the following functions.

```
a.  (1○α)*2
b.  (2○α)*2
c.  (1○¯3+4×α)*2
d.  (2○ω)*4
e.  ((SIN ω)*2)×(COS ω)*2
```

8.15 Verify the following sequence of identities:

```
((1○ω)*3)×(2○ω)*4
(1○ω)×(1-(2○ω)*2)×(2○ω)*4
(1○ω)×+/1 ¯1 1×(2○ω)*0 2 4
```

Use this sequence to show that an integral of the function `((1○ω)*3)×(2○ω)*4` is `-+/(1 ¯1 1÷1 3 5)×(2○ω)*1 3 4`.

8.16 Use the procedure illustrated in Ex 8.15 to produce an expression that defines an integral of each of the following functions.

```
a.  ((1○α)*5)×(2○α)*4
b.  ×/(1 2○α)*3 5
c.  ×/(5 6○ω)*4 5
d.  ((COS α)*4)×(SIN α)*5
```

8.17 Use the addition formula for sine to show that

```
(1○α)×2○ω
0.5×(1○α+ω)+1○α-ω
```

Use this identity to produce an integral of the function `(1○3×ω)×2○7×ω`.
Hint. Substitute `3×ω` for `α` and `7×ω` for `ω` in the above identity.

8.18 Use the identity in Ex 8.17 to produce an integral of each of the following functions.

a. (2○4×⍺)×1○6×⍺
b. (*SIN* 2×⍵)×*COS* ⍵
c. (2○¯3×⍵)×1○⍵
d. (2○4×⍺)×1○4×⍺

8.19 Use the addition formulas for the sine and cosine functions to establish identities from which integrals of the following functions can be produced. Use these identities to produce the integrals.

a. (*COS* 2×⍺)×*COS* 3×⍺
b. (*SIN* 3×⍵)×*SIN* 6×⍵
c. ×/1○4 9×⍵
d. ×/2○¯4 2×⍺

8.20 The rules described in Section 8.3 for producing integrals of certain functions can be used to produce the solutions of certain simple initial value problems. For example, consider the problem of determining the function *F* such that *D̲F* ⍵ equals (*F* ⍵)*2 and 2=*F* 1. Equivalently, (*D̲F* ⍵)÷(*F* ⍵)*2 equals 1. An integral of the function *ONE*:1+0×⍵ is ⍵, and an integral of (*D̲F* ⍵)÷(*F* ⍵)*2 is -÷*F* ⍵. Consequently there is a constant *S* for which ⍵+*S* equals -÷*F* ⍵ or equivalently *F* ⍵ equals -÷⍵+*S*. Since 2=*F* 1 we have *S*=¯0.5, so that *F*:-÷⍵-0.5. Use this procedure to determine the solution *F* of each initial value problem below.

a. *D̲F* ⍵
   ÷*F* ⍵

and 4=*F* 10

b. *D̲F* ⍺
   3+2×*F* ⍺

and ¯1.5=*F* 5

c. *D̲F* ⍺
   ÷0○⍺

and ¯4.5=*F* 0

d. *D̲F* ⍵
   ÷*COS F* ⍵

and 0=*F* 0.5

8.21 For each function and pair of scalars *A* and *B* below, determine the signed area of the region in a plane bordered by the graph of the function, the horizontal axis, and the vertical lines at *A* and *B*.

a. 2○⍺ ; *A*←-○÷2 ; *B*←○÷2
b. 2○⍺ ; *A*←0 ; *B*←○2
c. (1○2×⍺)*2 ; *A*←-○÷2 ; *B*←○÷2
d. (*SIN* ⍵)×*COS* 4×⍵ ; *A*←-○÷2 ; *B*←○÷2
e. ⍺×*E* ⍺*2 ; *A*←0; *B*←1
f. ÷1-⍵ ; *A*←2; *B*←3

8.22 Suppose that *H*:|(*F* ⍵)-*G* ⍵ and that *S̲H* is the definite integral of *H*. Then the function value *A S̲H B* is identical to the area of the region in a coordinate plane bordered by the graphs of *F* and *G* and the vertical lines at *A* and *B* (Ex 5.44). Determine the area of this region for each pair of functions and pair of scalars *A* and *B* below.

a. 1○⍺; 2○⍺; *A*←0; *B*←○÷2
b. 1-⍵*2; ⍵*2; *A*←0; *B*←1
c. *⍺; 3-*⍺; *A*←¯1; *B*←2

8.23 Suppose that *K* and *L* are positive integers and define *F*:(*SIN K*×ω)×*COS L*×ω. Use the identity in Ex 8.17 to show that 0=(○¯1)*SF* ○1, where *SF* is the definite integral of *F*.

8.24 Suppose that *K* and *L* are positive integers and define *F*:(*SIN K*×ω)×*SIN L*×ω and *G*:(*COS K*×ω)×*COS L*×ω. Use the identities derived in Ex 8.19 to show that the following propositions have the value 1 for the definite integral *SF* and *SG* of *F* and *G* respectively.

a. (*K*≠*L*)≤0=(○¯1)*SF* ○1
b. (*K*=*L*)≤(○1)=(○¯1)*SF* ○1
c. (*K*≠*L*)≤0=(○¯1)*SG* ○1
d. (*K*=*L*)≤(○1)=(○¯1)*SG* ○1

8.25 Use integration by parts to produce an expression that defines an integral of each of the following functions.

a. (α*2)××α Hint. (α*2)××α is identical to α×(α××α); use this identity and example 1 of Section 8.4.
b. ω×¯1○ω
c. ⍟ω Hint. See example 2 of Section 8.4.
d. ω×⍟ω
e. ¯2○*W*
f. ¯3○ω
g. ¯5○ω
h. ¯6○ω
i. ¯7○ω
j. α×*COS* 3×α
k. α×*ARCTAN* α
l. 1○⍟ω
m. (ω*2)÷0○ω
n. (*LN* ω)÷ω
o. α×*SIN* α

8.26 For each function *F* below use integration by parts to produce a function *IF* such that for each non-negative integer *N*, *N IF* ω is an integral of *N F* ω.

a. *F*:(ω*α)×2○α
b. *F*:(ω*α)×(*B*+*A*×ω)*0.5
c. *F*:(3○ω)*α
d. *F*:(⍟|ω)*α

e. Repeat this exercise for the function *F*:×/(1 2○ω)*α, where *N* is now a vector of length 2 whose elements are non-negative integers.

Hint. Apply integration by parts to both functions *G*:(*I*,α)*F* ω and *H*:(α,*J*)*F* ω for non-negative integers *I* and *J*.

8.27 Use the substitution method to produce an expression that defines an integral of each of the following functions.

a. (α+4)÷(α+2)*0.5 Hint. For the substitution method as described in Section 8.5, define *G*:¯2+α*2.
b. (1-ω*2)*0.5 Hint. Define *G*:2○ω.
c. ÷(1+ω*2)*0.5 Hint. Define *G*:5○ω.
d. α×(1+α)*÷3
e. (¯1+ω*2)*-0.5
f. ÷1+α*÷3
g. α×(¯4+α*2)*0.5
h. ÷ω×(9+ω*2)*0.5
i. ÷(4+α*2)*0.5
j. (ω*2)÷(9-ω*2)*0.5
k. α÷(¯1+2×α)*0.5

8.28 Show that for a constant *A* the function (÷*A*)×¯3○ω÷*A* is an integral of ÷(*A*⋆2)+ω⋆2 and that (÷*A*)×¯7○ω÷*A* is an integral of ÷(*A*⋆2)-ω⋆2. Using these results as a guide produce an expression that defines an integral of each of the following functions.

a. ÷9-α⋆2
b. ÷16+ω⋆2
c. ÷(16+ω⋆2)⋆0.5
d. ÷¯4+α⋆2

8.29 Show that for scalars *A* and *B*:

A+(*B*×ω)+ω⋆2
(*A*-(*B*÷2)⋆2)+(ω+*B*÷2)⋆2

A+(*B*×ω)-ω⋆2
(*A*+(*B*÷2)⋆2)-(ω-*B*÷2)⋆2

These identities are often referred to as completing the square.

8.30 The identities in Ex 8.29 are helpful in producing integrals of certain functions. For example, consider the function *F*:÷3+(4×ω)+ω⋆2. According to the first identity in Ex 8.29, *F* ω equals ÷¯1+(ω+2)⋆2, or equivalently *F* ω equals ÷¯4○ω+2. Consequently ¯6○ω+2 is an integral of *F* ω. Using this example as a guide, produce an integral of each of the following:

a. ÷¯8+(6×α)-α⋆2
b. α÷(5+(4×α)+α⋆2)⋆0.5
c. ω÷8+(2×ω)-ω⋆2
d. (3+2×α)÷5+(4×α)+4×α⋆2

8.31 Produce an expression for the definite integral function of the rational function (*B POLY* ω)÷*C POLY* ω for each pair of coefficient vectors *B* and *C* below:

a. *B*←1 0 1; *C*←0 ¯2 1 1
b. *B*←1 0 2; *C*←*CFR* 2 2 2
(see Ex 3.38 for the definition of *CFR*)
c. *B*←1 0 0 1; *C*←0 ¯4 0 1
d. *B*←1 3; *C*←¯4 0 1 *PPROD* ¯4 0 1
e. *E*. *B*←13 15 1 ¯2; *C*←4 2 ¯1

8.32 Recall the identity for bi-molecular reactions given in Ex 6.83, namely, *D̲N* ω equals *K*×(*A*-*N* ω)×*B*-*N* ω, where *K* and *A* and *B* are constants.

a. Show that

÷*K*×(*A*-α)×*B*-*A*
(÷*K*×*B*-*A*)×(÷*A*-α)-÷*B*-α

b. Verify the sequence of identities

1
(*D̲N* ω)÷*K*×(*A*-*N* ω)×*B*-*N* ω
(÷*K*×*B*-*A*)×((*D̲N* ω)÷*A*-*N* ω)-(*D̲N* ω)÷*B*-*N* ω

and consequently

```
K×B-A
((D̲N ω)÷A-N ω)-(D̲N ω)÷B-N ω
```

c. Show that there is a scalar *S* for which

```
S+ω×K×A-B
⍟(A-N ω)÷B-N ω
```

and if `0=N 0` then

```
ω×K×A-B
⍟(1-(N ω)÷A)÷1-(N ω)÷B
```

d. Show that

```
N ω
(A×B×1-*ω×K×A-B)÷B-A×*ω×K×A-B
```

e. Compare the values of *N* with the approximate values produced in Ex 6.83.

8.33 Suppose that `G:2×¯3○α` and show that

```
SIN G α                   COS G α
2×α÷1+α*2                  (1-α*2)÷1+α*2
```

Use these identities and the substitution method with `G:2×¯3○α` to produce an expression that an integral of each of the following functions.

a. `÷1+1○α`
b. `÷1-2○α`
c. `(COS ω)÷2-COS ω`

8.34 Use the procedure illustrated in Ex 8.20 to produce the solution *F* of each of the following initial value problems.

a. `D̲F ω` equals `(2-F ω)×3-F ω` and `0.5=F 0`.
b. `D̲F α` equals `2+COS F α` and `1=F 1`.

8.35 Produce an expression that defines an integral of each of the following functions.

a. `(*α)÷2+*2×α`
b. `(ARCCOS ω)÷0○ω`
c. `⍟(1-α)*0.5`
d. `(α*-0.5)×1○α*0.5`
e. `(COS ω)÷1+(SIN ω)*2`
f. `α×(1○α)*2`
g. `ω÷15+(8×ω)+ω*2`
h. `÷(1-*-α)*0.5`
i. `(ω*2)×COS ω`
j. `ω×COS ω*2`
k. `(α*2)×E α*3`
l. `(SIN 2×ω)÷1+COS ω`
m. `(ω*2)×¯3○ω`
n. `(α*3)×E α*3`
o. `÷9 8 ¯1 POLY ω`
p. `÷(9 8 ¯1 POLY ω)*0.5`
q. `(ARCTAN *α)÷*α`
r. `(ω*2)÷¯6+ω+ω*2`
s. `(COS 5×ω)×COS 4×ω`
t. `(3+α)÷(6×α)+α*2`
u. `÷ω+⍟ω`
v. `((4+3×α)÷4-3×α)*0.5`
w. `2×(1○α)*2`
x. `ω×3*ω*2`
y. `(¯5+2×α)÷8+(¯4×α)+α*2`
z. `(2○α)*5`

8.36 Sketch the graph of the function *F*:÷ω*0.5 and define its definite integral *S̲F*. Show that 0 is a valid argument of *S̲F*, even though it is not a valid argument of *F*. Zero is said to be an improper argument of *S̲F*.

8.37 a. Sketch the graph of the function *G*:⍟ω and define its definite integral *S̲G*.

b. Define *A*:÷ω and *B*:⍟ω. The values of both ÷*A X* and ÷*B X* approach 0 as *X* is assigned values close to 0. It is then true that the values approached by (*B X*)÷*A X* and (*D̲B X*)÷*D̲A X* as *X* is assigned values close to 0 are identical (this follows from a variation of L'Hospital's rule, Ex 8.2). Use this fact to show that the values of (*B X*)÷*A X* approach 0 as *X* is assigned values close to 0. Therefore 0 is a valid argument of *S̲G*, even though it is not a valid argument of *G*.

8.38 The Riemann zeta function *Z* is defined formally by defining each value *Z S* to be the value approached by the values ×/÷(1+ι*N*)**S* as *N* is assigned large values. According to Ex 7.52a, the values +/÷(1+ι*N*)*1 or equivalently ×/÷1+ι*N*, increase without bound as *N* is assigned large values and do not approach a value. Evidently this is also true of +/÷(1+ι*N*)**S* if *S*≤0. In part b below we will show that this is also true if (0<*S*)∧*S*<1. However, we will show in part a that these values do approach a value if *S*>1. Thus the domain of the Riemann zeta function is ω>1. The method used in this example is known as the integral test. The interest in the Riemann zeta function stems from its role in proving the Prime Number Theorem, which states that the values (*PR N*)÷*N*÷⍟*N* approach 1 as *N* is assigned large values, where the value *PR N* is the number of prime numbers not exceeding *N*.

a. Suppose that *S*>1 and consider the suffix +/*K*↑÷(1+ι*N*)**S*. Show that +/*K*↓÷(1+ι*N*)**S* is a Riemann sum of the function ω*-*S* for the partition *K*,*K*↓1+ι*N* and the vector *K*↓1+ι*N* whose elements separate those of the partition. Sketch the graph of ω*-*S* on the interval (*K*≤ω)∧ω≤*N* and, on the same graph, sketch the region composed of rectangles whose area is +/*K*↓÷(1+ι*N*)**S*. Use this graph and the definite integral of ω*-*S* to argue for the fact that if *EST*:(÷1-α)×ω*1-α then:

(*S EST K*)≥+/*K*↓(1+ι*N*)**S*

Note that for large values of *K* the values *S EST K* are small if and only if *S*>1. Arguing as in section 6.1, we see that if *S*>1 then the values +/÷(1+ι*N*)**S* approach some other value as *N* is assigned large values. Use the above inequality to produce the values *Z S* to within 3 decimal places for *S*←1.5; *S*←2; *S*←5.

b. Suppose that (0<S)∧S<1 and sketch the graph of ω⋆-S on the interval ω≥1. Show that +/÷(1+⍳N)⋆S is a Riemann sum of ω⋆-S for the partition 1+⍳N+1 and the vector 1+⍳N whose elements separate those of 1,1+⍳N, and on the same graph sketch the region composed of rectangles whose area is this Riemann Sum. Define the definite integral SF of F:ω⋆-S and use the graph to argue that

```
(1 SF N+1)≤+/÷(1+⍳N)⋆S
```

Finally, argue for the fact that if (0<S)∧S<1 then the values (1 SF N+1) (and consequently the values +/÷(1+⍳N⋆S)) increase without bound as N is assigned large values.

8.39 Use the Second Fundamental Theorem of the Calculus to produce the value of each of the following definite integrals.

```
a.  0 SH 1            ; H:ω×⍟ω
b.  1 SF 4            ; F:(α⋆2)×(3×α⋆3)⋆0.5
c.  1 SK 2            ; K:¯4○2×ω
d.  2 SF 5            ; F:÷¯12 5 2 POLY ω
e.  0 SL ○2           ; L:(⋆α)×2○α
f.  0 ST ○÷4          ; T:○(3○ω)⋆2
g.  (-○÷2) SH ○÷2     ; H:(ω⋆2)×¯1○ω
h.  2 SA 5            ; A:(ω⋆3)×(¯4+ω⋆2)⋆0.5
i.  0 SF ○1           ; F:(1○α)÷1+(1○α)+2○α
j.  (÷4) SB 3÷4       ; B:÷ω×0○ω
k.  0 SC 1            ; C:(ω⋆¯0.5)×1 3 5 POLY ω
l.  0 SF 1            ; F:÷1+ω+ω⋆2
m.  1 SH 5            ; ω×⋆1+ω⋆2
n.  0 SC ○1           ; C:(2○ω)÷(1○ω)⋆0.5
o.  ¯1 SH 1           ; H:((1+ω)÷1-ω)⋆0.5
```

Appendix A

The following is a summary of the detailed development in Chapter 10 of Iverson [2]. The indented expressions must be entered at a computer keyboard exactly as they appear. An expression at the margin is the value of the preceding indented expression and is produced by the computer. These values serve as checks on the entered expressions. You must receive the correct value at each step before proceeding further.

```
      )CLEAR
CLEAR WS
      ⎕IO←0
      ⎕FX 2 8ρ'Z←SUB9 YZ←Y
SUB9
      M←2 34ρ(34↑'ON9'),'(2,M9)ρ(M9↑X),(M9←(+/X=X)⌈+/Y=Y)↑Y'
      AR9←3 7ρ'   Z←    Y Z←   Y Z←X
      ⎕FX 3⌽AR9[2 0;],M
ON9
      G9←'XY'
      M←'CAN9' ON9 '3⌽AR9[1 0×+/⌈/G9∘.=Y;],X ON9 SUB9 Y'
      ⎕FX 3⌽AR9[2 0;],M
CAN9
      ⎕FX 'IS9' CAN9 '⎕FX X CAN9 Y'
IS9
      H9←':'
      'P9' IS9 '(+/⌊\Y≠H9)↑Y'
P9
      'S9' IS9 '(1++/⌊\Y≠H9)↓Y'
S9
      'SDEF9' IS9 '(P9 Y) IS9 S9 Y'
SDEF9
      A9←'⍎((0='
      B9←')/'''
      C9←'''),(0≠'
      D9←')/'''
      E9←''''
      M←'FORM9:A9,(P9 S9 S9 Y),B9,(P9 S9 Y),C9'
      M←M,',(P9 S9 S9 Y),D9,(S9 S9 S9 Y),E9'
      SDEF9 M
FORM9
      SDEF9 'RDEF9:(P9 Y) IS9 FORM9 Y'
RDEF9
      RDEF9 'ADEF9:SDEF9 Y:3=+/H9=Y:RDEF9 Y'
ADEF9
      C←'IN9:X IN9 (Q9↑Y),(1↓X),(1+Q9)↓Y'
      C←C,':(+/Y=Y)=Q9←+/⌊\Y≠1↑X:Y'
      ADEF9 C
IN9
```

```
      I9←'α X9 '
      J9←'ω Y9 '
      K9←'αω'
      ADEF9 'SW9:K9 IN9 Y:1≠+/⌈/K9∘.=Y:Y'
SW9
      ADEF9 'SUB9:I9 IN9 J9 IN9 SW9 Y'
SUB9
      AR9←3 9ρ'   Z9←    Y9Z9←    Y9Z9←X9 '
      G9←'αω'
      ADEF9 'DEF:0ρADEF9 ⍞'
DEF
```

The following changes in the definition of the function *DEF* permit function definitions in the αω form to be displayed at a computer. This feature is helpful for reviewing function definitions, and will be used in Chapter 8 to produce derivatives at a computer.

```
      F9←' '
      R9←'R̲9'
      T9←'←Y9'

      ADEF9 'STL9:(~∧\F9=ω)/ω'
STL9
      ADEF9 'REF9:⍎R9,(STL9 (ωιH9)↑ω),T9'
REF9
      ADEF9 'BDEF9:ADEF9 REF9 ω'
BDEF9
      ADEF9 'CDEF9:⍎R9,STL9 ω:3≠⎕NC ω:0ρω'
CDEF9
      ADEF9 'DDEF9:0ρBDEF9 ω:0=+/H9=ω:CDEF9 ω'
DDEF9
      ADEF9 'DEF:DDEF9 ⍞'
DEF
```

References

1. Iverson, K.E., Algebra: an algorithmic treatment, Addison-Wesley, Reading, Mass., 1972.

2. Iverson, K.E., Elementary Analysis, APL Press, Swarthmore, Pa., 1976.

Index